THE PREDICTION OF IONOSPHERIC CONDITIONS

GEOPHYSICS AND ASTROPHYSICS MONOGRAPHS

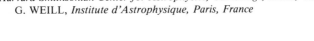

THE PREDICTION OF IONOSPHERIC CONDITIONS

by

G. S. IVANOV-KHOLODNY

and

A.V. MIKHAILOV

Institute of Applied Geophysics, Goscohydromet, Moscow, U.S.S.R.

D. REIDEL PUBLISHING COMPANY

A MEMBER OF THE KLUWER ACADEMIC PUBLISHERS GROUP

DORDRECHT / BOSTON / LANCASTER / TOKYO

Library of Congress Cataloging in Publication Data

Ivanov-Kholodny, G. S. (Gor Semenovich)
 The prediction of ionospheric conditions.

 (Geophysics and astrophysics monographs)
 Translation of: Prognozirovanie sostoíaniía ionosfery.
  Bibliography: p.
  Includes index.
 1. Ionospheric forecasting. 2. F region. I. Mikhaılov, A. V. (Andreı
Valer'evich) II. Title. III. Series.
 QC881.2.I6I9313   1986   551.5′145   86–13960
 ISBN 90–277–2143–2

Published by D. Reidel Publishing Company,
P.O. Box 17, 3300 AA Dordrecht, Holland.

Sold and distributed in the U.S.A. and Canada
by Kluwer Academic Publishers,
101 Philip Drive, Assinippi Park, Norwell, MA 02061, U.S.A.

In all other countries, sold and distributed
by Kluwer Academic Publishers Group,
P.O. Box 322, 3300 AH Dordrecht, Holland.

Originally published in Russian by Gidrometeoizdat:
прогнозирования состояния ионос феры
Translated by J. Schmorak
Original © Gidrometeoizdat

Printed in The Netherlands

TABLE OF CONTENTS

FOREWORD

The ionosphere of the Earth has been actively studied since the 1920's, following the discovery of ground radio-sounding. By means of this method results were obtained by an international network of ionospheric stations, in particular, by the successful implementation of a number of rigorously planned international scientific research programs,* enabling the collection of extensive experimental material on some of the most important parameters of the ionosphere – the critical frequencies of E-, $F1$- and $F2$-layers. Comprehensive analyses of these observation data give a fairly complete picture of the various changes taking place in the principal ionospheric layers at different points on our globe.

Another important aspect of the study of the ionosphere, which has been in progress for the past three decades, is an extensive program of *in situ* determinations of the various physical parameters – first using rockets, and subsequently artificial satellites. The data thus obtained on the principal ionizing agent – short-wave solar radiation – and on the physical conditions prevailing in the upper atmosphere and in the ionosphere at different altitudes, allow the proposal of a self-consistent mechanism of ionosphere formation. A general outline of the theory of ionosphere formation at different altitudes is now complete.

Its application to specific cases, dependent on a more accurate determination of input parameters to give solutions valid for a definite set of conditions etc., is yet to be accomplished.

The use of artificial satellites in cosmic research yielded abundant scientific data. Most important, satellites yield almost instantaneous latitudinal sections of the upper atmosphere and the processing of such data resulted in semi-empirical models of planetary distributions of various parameters of the upper atmosphere for the first time. Input data include: the solar activity index $F_{10.7}$, and indexes of geomagnetic activity, A_p or K_p, since the state of the upper atmosphere is solely determined by short-wave radiation and corpuscular solar radiation.

A planetary description of the ionosphere, with its numerous parameters, and of its evolution with respect to time proved much more difficult than for the neutral atmosphere. This explains why empirical models of planetary distribution have so far only been developed for the critical frequencies of the principal ionospheric layers. However, the complete planetary picture of the height distribution of the principal ionospheric parameters is required for geophysical studies.

---

* International Polar Year (1932–1933); International Geophysical Year and International Geodesic Collaboration Year (1957–1959); International Year of the Quiet Sun (IQSY, 1964–1965). The International Year of Maximum Solar Activity (1979–1981) began in August 1979.

Prediction of ionospheric conditions is essential in radiocommunication, which involves the propagation of radio waves through the ionosphere. This particular problem was presented and solved in several different ways at various stages of ionospheric research. Such predictions were based mostly on empirical data on the quality of radio-wave communication along different paths, and on morphological information about the behavior of the ionosphere (empirical approach). As the principal mechanisms of ionosphere formation became better understood, and as more sophisticated computers became available, in certain cases such predictions could be treated as physical problems of computing the state of the ionosphere under specific conditions (deterministic approach). This approach and its possible applications at the present time are discussed in Chapter 5.

Deterministic prediction of the state of the ionosphere is possible, since the state of the upper atmosphere and conditions of ionization are determined by the short-wave and corpuscular radiation emitted by the Sun. Thus our primary objective is twofold: 1) to carry out a comprehensive study of the physical mechanisms which determine this state and develop the techniques for the incorporation of these effects in predictions and 2) to predict the geophysically effective solar radiations from observed solar activity data. The first part of this task is treated in more detail in the chapters that follow; the possibility of predicting short-wave radiations is discussed in Chapter 3.

Computing the state of the ionsphere under the different conditions is a complex task indeed. Accordingly, the deterministic approach is only of interest if we derive a program for calculating ionospheric parameters which is realizable in practice, allows for the complexity of the processes responsible for the formation of the ionosphere, and yields sufficiently accurate results. This problem has not yet been completely solved, but under certain conditions (in the daytime, at middle latitudes) it is possible to use recent scientific results to calculate the principal ionospheric parameters of the $F2$-region theoretically with reasonable accuracy. The critical frequencies calculated by this method should be at least as accurate as those obtained by the empirical methods employed at present. The method for calculating the periodic variations of the $F2$-region developed by the authors is presented in Chapter 4.

As has been pointed out, one of the pertinent problems of the present day is to develop the general theory of ionosphere formation to the point where it may be used in operational computations. This problem may be tackled in different ways, since different aeronomic parameters, including data on short-wave solar radiation, the parameters of the neutral atmosphere and reaction constants, may be selected as the input magnitudes. Our approach to the selection of the fundamental parameters is explained in Chapter 3. A discussion of input parameters and general assumptions underlying the theory of ionosphere formation, of secondary importance for our purposes, is given in Chapter 2. For more complete and detailed treatment of these problems the reader is referred to a number of recent monographs included in the list of references.

Chapter 1, which is introductory, contains a statement of our objectives, a discussion of the principal problems and aims of ionospheric predition, and a description of the current methods of short-term and long-term predictions, based on empirical and semiempirical ionospheric models.

Several attempts have been made in the past to calculate the physical parameters of the ionosphere using the theory, by solving appropriate theoretical equations for the ionosphere formation. In each case the latest advances in the theory of ionosphere formation were included, the most sophisticated computing machines available were employed, and the most recent values of the aeronomic parameters were used in computation. The results of successive calculations improved accordingly. We also adopted the same approach. We believe that the latest advances in the field have changed the situation so radically that the development of a deterministic method of predicting the state of the ionosphere based on semi-empirical models of the upper atmosphere and on the feasibility of predicting relative variations of short-wave solar radiation (Chapter 3) has now entered the realm of possibility. This not only applies to the $F2$-region, which is the subject of this book, but also to other ionospheric regions (for the E-region see /35/). Many problems are common to different aspects of ionospheric prediction (critical frequencies of E-, $F1$- and $F2$-layers, maximum usable frequencies (MUF), lower usable frequencies (LUF), propagation of radio waves, etc.). The subject of this book is the $F2$-layer – the principal ionospheric layer, which also determines the long-range propagation of decametric radio waves.

Attempts to make accurate evaluations of ionospheric parameters under specific helio-geophysical conditions revealed that such calculations are often difficult and unreliable owing to inaccurate values of the input parameters selected, their mutual inconsistency, and owing to the fact that reliable, simultaneously determined values are not available. Many questions had to be reconsidered from first principles.

The problem has proved to be so complex, and the available experimental data so abundant that a reasonably complete presentation in a single article or even in a series of articles is clearly impossible. For this reason, the principal objective of this monograph is to present the reader with the method developed by the the authors for computation and prediction of the state of the $F2$-layer of the ionosphere.

It is our pleasant duty to express our gratitude to Professor N.P. Ben'kova who offered several valuable comments on the manuscript; to G.I. Ostrovskii and B.E. Serebryakov for their help with the computations, and to N.I. Selyakova and E.N. Torpishchina for their aid in editing and preparing the manuscript for printing.

CHAPTER 1. IONOSPHERIC PREDICTION – PRESENT-DAY PROBLEMS AND APPROACHES

1.1 IONOSPHERIC PREDICTION

The advances in research since the advent of rockets and satellites, have in fact enhanced rather than diminished the importance of ionospheric radiocommunication. The data yielded by *in situ* measurements made in rockets and satellites have substantially extended and improved our knowledge of the ionosphere, in particular with respect to the mechanism of formation. It is now possible to use the latest theoretical and experimental advances to develop a deterministic approach to ionospheric prediction and to calculate the state of the ionosphere under specific helio-geophysical conditions.

Before enumerating the problems in predicting the state of the ionosphere, we shall explain the meaning of certain terms and give some fundamental definitions.

1.1.1 Terminology. "Prediction" will be understood as the forecast of changes in a specific phenomenon to be expected in the future.

Predictions are less valuable than accurate computations based on firmly estab-lished laws (e.g., computing the times of solar eclipses in celestial mechanics). On the other hand, they are more valuable than non-scientific guesses, guided by intuition rather than by intelligent application of the laws of nature.

Thus, predictions are the last resort when accurate computations such as would be acceptable in the natural sciences, are impossible. They are used whenever the information available is incomplete, inaccurate, unreliable etc. The degree of the accuracy of a prediction is the greater, the greater the extent to which it is based on deterministic calculations.

The information about a given phenomenon may become so complete and accurate that in the course of time predictive estimates are replaced by deterministic calculations with accurate estimates of the errors involved. In our view, the knowledge now available about certain ionospheric processes is actually approaching such a transition stage, and we shall attempt to demonstrate this point where applicable. However, other relationships governing the evolution of the ionosphere, involving unpredictable sporadic disturbances have not as yet been sufficiently investigated. Thus, accurate computations can only be conducted for moderate latitudes, during quiet periods, while we have to make do with statistical probability predictions, which are less accurate, for perturbed periods.

There are two basic approaches to prediction: the deterministic and the statistical. The latter approach is based on morphological laws (qualitative and quantitative) obtained by statistical processing of data obtained by observations, and therefore a definite probability can be assigned to each prediction.

The deterministic approach to prediction is based on rigidly quantitative descriptions of physical effects and relationships involved in the process. As applied to the ionosphere, this means that it is based on a sufficiently reliable theory of ionosphere formation and evolution with respect to solar and geophysical factors.

In addition to this method, there are also the inertial and the intuitive methods of prediction.

The inertial method is based on the principle "what happens today, should also happen tomorrow", since the period during which any given change takes place is usually much longer than one day. Thus, for instance, for certain weather phenomena, this period is about three days, which means that the probability that an inertial prediction will be correct is quite high ($\frac{2}{3}$). The acccuracy of this method in predicting short-wave solar radiation, outside the periods of solar flares, is even more reliable, since the typical period here is even longer (about 5 days). This means that even the use of a single inertial criterion for prediction of an effect may sometimes be quite effective. In this sense, inertial prediction is a particular case of statistical probability prediction. If the tendency of the effect to change is allowed for in the prediction, the method is termed quasi-inertial.

In the intuitive method of prediction – essentially a subjective one – certain relationships are used to reach the final conclusions intuitively either by experienced or by lucky forecasters. In this book only formal predictions, which do not depend on the personality of the forecaster, will be considered.

1.1.2 Ionospheric Parameters. In order to reach a general characterization of the problems involved in the prediction of the ionosphere, we shall consider the factors determining the state of the ionosphere and explain why it is impossible to produce a fully deterministic calculation. We shall consider a) the fundamental parameters of the ionosphere; b) their spatial and temporal distribution; and c) their dependence on solar and geomagnetic activities.

a) The state of the ionosphere is determined by a rather large number of different parameters. The characteristics of the ionized fraction of the upper atmosphere such as electron concentration, ion composition, electron and ion temperatures, particle fluxes and drifts, constitute the basic physical parameters of the ionosphere. It may be necessary to study a self-consistent problem while simultaneously allowing for the changes in the state of the ionosphere and the neutral upper atmosphere. However, in this book, in considering ionospheric changes we shall assume that the parameters of the neutral atmosphere are fixed.

b) It should be noted, first of all, that the parameters of the ionosphere undergo considerable changes with altitude. The concentrations of ions, n_i, and electrons n_e, at certain altitudes have characteristic maximum peaks, forming the ionospheric layers, E, $F1$ and $F2^*$. As a result, the stratified structure of the ionosphere may be characterized by a set of parameters: the concentration n_e^m at layer maximum, layer

---

* We shall use the term "ionospheric region"; when discussing the ionosphere throughout the altitude range in which it displays certain physical properties. The ionospheric region close to the peak of n_e will be referred to as the "ionospheric layer".

altitudes and widths, the depths and widths of the troughs between layers etc. This group of parameters describes the special features of the distribution of n_c over the altitude h or, as it is usually referred to, the profile, $n_c(h)$.

In addition to variation with the altitude, ionospheric parameters also vary with the geographical coordinates (latitude, φ_1, longitude, λ), and with the geomagnetic coordinates. The properties of the ionosphere are known to undergo significant changes at the equatorial and auroral latitudes. Accordingly, we may distinguish between mid-latitude, equatorial and auroral ionospheres. In this book, only the ionosphere at moderate latitudes will be considered.

Knowledge of the relationship governing the variation of the ionosphere with time is particularly important in prediction. Thus, intensive efforts have been made to determine the special features of diurnal variations. It was found that there were variations in parameters not only between daytime and nighttime, but also that there were sharp variations between values for the daytime, even around noontime, from one place to another. Indeed there are special features which are solely a function of the latitude. It was also found that, for a given site, the diurnal variations of the ionosphere depend on the season of the year, giving rise to the so-called seasonal and December anomalies. Thus, any description of the variation of the ionosphere with the time includes several parameters.

c) Certain variations in the ionosphere were found to be caused by the variation of active factors such as solar radiation and geomagnetic activity. Therefore, the ionosphere is not merely a geophysical, but also a helio-geophysical entity.

Owing to the abundance of the parameters required to describe it, and since each parameter varies differently with the coordinates, time, and active factors, the behavior of the ionosphere is extremely complex (cf. para. 1.2). As a result, both the study of the ionosphere and its description are also very complex tasks.

Nevertheless, as a result of the successful realization of extensive studies of the ionosphere from the ground, and from rockets and satellites, we have now advanced to the point where we can deal not only with statistical predictions of the variations of individual parameters on the planetary scale, but also with deterministic forecasting of such variations. In this book we intend to show that advances in this direction are very impressive indeed.

1.1.3 Objectives. Ideally, any method of ionospheric prediction should enable calculation of all the ionospheric parameters and their planetary and altitude distributions at any future point in time, with the requisite detail and accuracy. In actual fact, the present-day "state-of-the-art" does impose a number of limitations, which we will proceed to discuss in some detail.

1. We must note, first of all, that such limitations may either be due to incomplete information, which might become more complete at some future date or they may be intrinsic to the phenomenon itself. For example, how detailed should the ionospheric charts of the different parameters be? This clearly depends on the accuracy with which each parameter can be determined, and on the extent of its variation from one point to another. Consider, for example, the maximum electron concentration, n_c^m, in the F-layer. We may ignore the minor irregularities, for which $\delta n_c/n_c \leq 10^{-2}$, and estimate the maximum regular latitudinal and longitudinal variations of n_c^m. The largest latitudinal gradient of n_c^m is found near the auroral and in the equatorial

zones, where it may be up to 10% over a distance of 2° lat. The largest longitudinal gradient of n_e^m is noted in the morning, and may be as large as 10% for two points 2–3° apart. At moderate latitudes, on the other hand, 10% gradients of n_e^m are noted over distances of 5–10° latitude and 20–50° longitude. These considerations determine the detail of the ionospheric charts in various parts of the globe required for determining n_e^m to a given degree of accuracy.

2. How detailed should the description of the parameters as a function of time be? This may be deduced from the above discussion bearing in mind that a change of longitude of 15° corresponds to a one-hour time difference. Clearly if only 10% accuracy is required in describing n_e^m in the $F2$-region, it is sufficient to take observations every hour, or even every $1\frac{1}{2}$ – 2 hours, except at sunrise and sunset. Furthermore, the extent to which the value of the median f_oF2, obtained from data taken over a period of a month, is a valid description of current values under quiet conditions in the ionosphere will be discussed in Chapter 5. Similar considerations apply to n_e-data as a function of the altitude, as well as to the other parameters.

3. The planetary distribution of ionospheric parameters was first studied solely using the data furnished by the ionospheric stations all over the world. Satellites, especially those with polar orbits, subsequently proved to a powerful tool in obtaining a global chart of the ionosphere. However, the relationships governing the global distribution of ionospheric parameters as a whole are still not completely understood.

4. Difficulties are also associated with the description of diurnal variations. This applies in particular to nighttime conditions, for which there is no generally accepted mechanism of formation of the $F2$ region. As long as the reason for the increase in electron concentration, noted on certain nights, is unknown, these predictions can only be approximate, and will be much less accurate than for daytime conditions.

5. Our incomplete knowledge of the $F2$ region in its perturbed state also imposes very serious limitations, not only on the prediction, but even on the understanding of the phenomenon itself. We are still unable to explain all the complexities of the evolution of the perturbation effect in space and in time. Moreover, except for the onset of recurrent geomagnetic disturbances, there is as yet no way of predicting the temporal variation of disturbances more than a few hours ahead.

6. As with other types of prediction, the nature and reliability of ionospheric prediction not only depends on an accurate and reliable knowledge of the relationships governing the evolution of the ionosphere, but also on the accuracy and the completeness of empirical and other information which is used in predictive computations. As we have pointed out, the prediction of data concerning geomagnetic perturbations is of very limited value; nevertheless, such data are used as the input data in all atmospheric predictions. A highly accurate prediction of short-wave radiation is only possible one or two days ahead. Long-range predictions – several months ahead – are based on the grosser estimates of average short-wave solar radiation, which merely yield median MUF frequency values.

Our limited knowledge of the ionosphere also causes difficulties in developing prediction methods. Many of these not only affect specific methods, but also the prediction as such. Some workers omit any mention of these difficulties, and for this reason their methods may appear more general and more effective than they really are. In deterministic methods of ionospheric prediction all such limitations should be clearly stated.

1.1.4 Modern Prediction Methods. We shall now proceed to give a general description of the prediction methods which are commonly employed. Quantitative predictions of the state of the ionosphere are made either by extrapolation or by the use of models. In the extrapolation method the future evolution of any given parameter is estimated from a number of observations made during a period of time. The magnitude of the parameter at some future point in time is estimated on the assumption that the trend of the evolution will remain unchanged in the future (quasi-inertial prediction). If models are used in prediction, it is possible to compute an entire set of parameters for the object of the model. If the state of the object is solely determined by certain conditions or forces, the task of prediction will consist of two stages: 1) prediction of the evolution of these conditions or forces; and 2) description of the state of the object under the predicted conditions.

In the prediction of the ionosphere, the fact that the condition of the ionosphere is chiefly determined by the solar activity level and by the level of geomagnetic perturbations which may be considered as a manifestation of the solar activity, is usually taken into consideration. Accordingly, ignoring the problem of predicting solar activity for the moment, the task of ionospheric prediction consists of relating the state of the ionosphere to the parameters of the solar activity, and of furnishing a more or less detailed description of ionospheric parameters. The ionospheric models, which are averaged for this purpose, and the actual long-term predictions of *median* parameter values, as well as short-term predictions of deviations from these parameters, will be discussed in paras. 1.2 and 1.3. The use made of the computation of ionospheric parameters for *specific* geophysical conditions in the prediction (the deterministic approach) will be described in Chapter 5.

Ionospheric variations can also be predicted independently of the solar activity, merely from a series of observations carried out during a previous period of time. This method is employed in both short-term and long-term predictions.[*]

The obvious question is the extent of validity of the inertial prediction method as applied to the ionosphere and how far ahead such an extrapolation may be made with adequate accuracy. These questions were discussed by Lavrova /46/, who based her arguments on the ionospheric data obtained at the Moscow station in the years of maximum (1969) and minimum (1964) solar activity. By computing the auto-correlations in a series of $\delta f_o F2$ (deviations from the median) it was shown that a significant daytime correlation is obtained only ½– 1 day ahead, while after 2, 3 and 27 days there is virtually no correlation. It was found that on subsequent days the value and the sign of $\delta f_o F2$ remain constant to within 0–15% in 90% of the cases under quiet daytime conditions, and in 85% of the cases in all phases of the solar cycle. If the ionospheric perturbation increases in the days that follow, the probability of the perturbed conditions increases as well, especially at night. Examination of the data taken at half-day intervals revealed that it is much more probable for a

[*] See, for example, C.M. Minnis, G.H. Bazzard).–J. Atmosph. and Terr. Phys., 1959, vol. 14, N 213; 1959, vol. 17, N 57; 1960, vol. 18, N 297.

perturbed night to follow a perturbed day than *vice versa*. Thus, the quasi-inertial method of prediction is only valid under quiet conditions.

Many types of prediction of ionopheric parameters are merely qualitative. They yield the approximate period of perturbation, and the perturbation intensity is estimated using a 7-point marking scale. Attempts were made to effect a quantitative prediction of f_oF2 by extrapolating the previous series of observations in some manner. Lyakhova and Kostina /49/ presented the results of an experimental prediction of this kind made at a numer of stations during a period of one year. The δf_oF2-values were predicted by extrapolating the previously observed values of this magnitude graphically, or by using the formula:

$$y = \sum_i a_i y_i,$$

where y_i are the current values of δf_oF2 obtained at hour intervals and a_i are the coefficients, which depend on the season of the year, the level and the sign of ionospheric disturbance, and on the time of the day at a particular station as well as on the period of time for which extrapolation is desired.

It was found that if three successively observed values of δf_oF2 are employed, it is possible to predict up to four hours ahead with a mean square deviation of 10–15%. Since n_e is proportional to f_o^2, a $\pm15\%$ error in f_o is equivalent to a deviation of 0.11 in log n_e (since, lg $(1 + \dfrac{\delta n_e}{n_e})$ = lg 1.3 = 0.11). Attempts to increase the prediction period proved unsuccessful.

Finally, Vsekhsvyatskaya et al. /9/ discussed the feasibility of expressing variations in the deviations of the critical frequencies δf_oF2 from the median values under various helio-geophysical conditions by specific relationships, which can then be used in making the predictions. It was found that the changes of f_oF2 due to random variations in the short-wave and corpuscular radiations of the Sun may be described by means of a stationary Poisson probability process with a probability of 85–95%. The random nature of the distribution of δf_oF2 confirms that the variations of f_oF2 can be correctly described by the median.

1.2 PREDICTION BASED ON IONOSPHERIC MODELS

One possible method of obtaining prediction algorithms is the use of ionospheric models, which permit temporal extrapolation of the formation of the ionospheric state. Depending on the task in hand, the models may differ in the number of parameters involved and in the principles of the construction; they may either be local or planetary. The construction of models involves the use of certain well-established physical relationships.

Thus, the quasi-inertial method of prediction is based on the assumption that the characteristic time of the processes to be forecast is longer than the period of the prediction. The method will now be used to establish a semi-diurnal ionospheric prediction. Thus, if the state of the ionosphere during the first 12 hours of the 24-hour period in question is the same as during the same period 24 hours earlier, it is assumed that the conditions of ionosphere formation remain stable, and the parameters during the following 12 hours will be the same as they were the day before. In using statistical models based on a large number of observations, it is assumed that under similar helio-geophysical conditions the average response of the ionosphere will be the same in the future as it was in the past. An example of this type of model is the "Prediction of Maximum Usable Frequencies" /73/, which is routinely employed in the Soviet Union in long-term prediction.

It should be noted that ionospheric prediction is presently based on averaged empirical models. The restricted nature of these models is due to the fact that they yield an averge picture of the state of the ionosphere, and consequently may only be employed for the prediction of median values. Specific helio-geophysical conditions on a given day are computed by other methods.

Ionospheric models may be empirical (statistical), semi-empirical or theoretical.

Empirical models are based on the generalization of a number of observations of individual ionospheric parameters, which may be presented in graphs, tables or formulas. Thus, statistical models /73/ and /4/ involve charts of planetary distribution of ionospheric parameters as a function of the coordinates, the local time and the level of solar activity. These models will be analyzed in Chapter 5. An example of a statistical model of another type is the "International Reference Model of the Ionosphere" /237/, which is based on experimental data. It permits the calculation of electron concentration profiles, the relative concentrations of various types of positive ions, profiles of the temperatures, T_e, T_i and T_n, and of collision frequency, and also yields information about the negative ions in the D-region. This model applies in certain latitudinal zones, in three seasons (winter, summer, equinox) and at three levels of solar activity.

We shall now discuss the difficulties involved in constructing a sufficiently complete statistical model of the ionosphere. The statistical models of the neutral atmosphere established in recent years, such as CIRA (1972), JACCHIA models, OGO-6 and MSIS would at first sight seem to indicate a way of solving the problem for the ionosphere as well. However, as was pointed out by Nisbet /225/, there is a fundamental difference between the two cases. The principal difficulty in constructing an ionospheric model is that a very large number of different parameters must be considered. Moreover, the complex effect of the geomagnetic layer on the ionospheric plasma must also be allowed for. However, there were too few observation data for the model of this type to be constructed.

Tables 1.1 and 1.2, taken from /225/, are a comparison of the available data against the necessary information for the construction of two types of models. Even if only 10 parameters are taken (n_e and the altitude of each ionospheric region), which is generally insufficient for a description of the state of the ionosphere, it is apparent that the amount of the information available is much less than would be required to construct an empirical model (e.g., with 10 possible levels allocated to each of the 7 independent variables).

The reason for the small amount of the information available as compared to that required for a statistical approach to the problem is due to the fact the stations conducting such observations are very unevenly distributed over the surface of the globe, and are altogether absent in several extensive areas. Observations involving external sounding of the ionosphere are still few in number, so that the selection of geophysical conditions is very limited.

TABLE 1.1

The required and the available information for the construction of an empirical model of the ionosphere

Region	No. of parameters	Information, bits	
		required	available
D	2	10^7	$6 \cdot 10^2$
E	2	10^7	$1 \cdot 10^5$
F1	2	10^7	$1 \cdot 10^5$
F2	2	10^7	$2,8 \cdot 10^5$
Above the maximum of F2 region	2	10^7	$9 \cdot 10^4$
Total	10	$5 \cdot 10^7$	$5,8 \cdot 10^5$

TABLE 1.2

The required and the available information for the construction of a theoretical model of the ionosphere

Parameters	Required information, bits	Estimated quantity of available information, bits
Thermosphere	512	374
Mesosphere	200	100
Flux of short-wave solar radiation	256	192
Collision and absorption cross-sections	1024	768
Winds	256	192
Electric fields	512	384
Reaction rate constants	400	300
Corpuscular fluxes	400	200
Magnetic Field	256	256
Total	3816	2766

It is apparent from Table 1.2 that when analytical relationships are used to describe parameter distribution, the establishment of a theoretical model becomes much easier. Only about 1000 bits of information are missing, as compared to $5 \cdot 10^7$ for the statistical model, so that in the former case it is more realistic to expect that the missing information will become available in the foreseeable future. Thus, the establishment of a theoretical model of the ionosphere seems to be much more promising using experimental data rather than by direct attempts to construct more complete empirical models.

It is clear that the principal obstacle in constructing complete statistical models is the requirement for extensive and comprehensive information about a large number of ionospheric parameters – electron concentrations, layer altitudes, ion composition, plasma temperatures etc., all of which require diversified and time-consuming observations. With the theoretical approach the principal difficulty is the selection of a self-consistent system of aeronomic parameters and, in addition, under perturbed conditions, the gathering of sufficient data on electric fields and corpuscular fluxes.

One possible way out of this dilemma is to develop semi-empirical models, which combine the features of both statistical and theoretical models. The theoretical part of such models is a system of differential equations, describing the effect of the main processes taking place in a given region of the ionosphere. The calculated distribution of the electron concentration may then be corrected by taking into account the altitude and the electron concentration in the $F2$-layer maximum, these values having been obtained by statistical processing of ionospheric sounding data. Such an approach eliminates some uncertainty as to the exact values of certain aeronomic parameters of the system, in particular, neutral composition, thermospheric winds and electric fields, which strongly affect the value of the altitude, h_m, and hence also the electron concentration, n_e^m. Examples of semi-empirical models are those used for the computation of the electron and ion concentration profiles at moderate latitudes /63,224/ as a function of the coordinates, the season of the year, local time and the level of solar activity (cf. Chapter 5).

It should be noted that semi-empirical ionospheric models seem to be the best suited for practical work – for example, in radio-wave ray tracing computations. In fact, the theoretical part of the model is usually simple, with the result that a profile of the electron concentration with the desired step size along the trajectory can be obtained within acceptable computer time. Since the profile is subsequently corrected, there is no need, for example, to solve for the diurnal variation of the electron concentration so that the time intervals taken may be fairly large. Further calculation then yields the profile of eletron concentration. This profile and its derivatives are continuous functions, which is important in solving the problems of radio wave propagation.* Thus, in routine calculations of ionospheric ray tracing the quality of the results may be improved if the critical frequencies of ionospheric layers are not used alone, but in conjunction with $n_e(h)$ profiles, calculated from semi-empirical models /66/.

Unlike the empirical models, the theoretical models take into account the principal physical mechanisms of ionosphere formation. The values of ionospheric parameters are obtained by solving a set of equations describing the principal physical laws as applied to the conditions prevailing in the upper atmosphere. The input parameters are: the intensity of short-wave solar radiations and the geomagnetic perturbation indexes (K_p, A_p), which describe the expected helio-geophysical environment. Sophisticated computers are required to obtain the results of the calculations within an acceptably short period of time. By developing suitable programs it is possible to obtain all the required ionospheric parameters by calculation. The theoretical foundations of the deterministic calculations are now sufficiently strong, but due to the lack of reliable values of the aeronomic input parameters, satisfactory computation of the $F2$ region has not as yet been carried out.

The theoretical model includes continuity equations and equations of motion to

---

*Problems of the representation of electron concentration profiles by models, and a brief characterization of ionospheric models as applied to radiocommunications, are discussed in the review /47/.

calculate the velocities of the thermospheric winds for electrons and for positive ions, and also thermal balance equations for the computation of the plasma temperatures. The most common approach to this problem today is to assume that the neutral composition and the temperature are given by a thermospheric model, but attempts are also being made to develop models in which the neutral composition, the temperature, and the concentrations of charged particles are calculated simultaneously by solving appropriate equations for neutral and ionized components of the atmosphere /41/.

In the context of the deterministic approach to ionospheric prediction, we must mention the work of Stubbe, who presented the most complete and detailed review thus far of a number of the fundamental problems involved in developing theoretical models, and was one of the first workers to point out the potentialities of this approach.

One of the models proposed by Stubbe /277/ may be taken as an example, which is of particular interest since it corresponds with the proposal made in this book, which is assumed to be self-consistent. During the preliminary stage, ionospheric reference data for n_c^m and h_m are used to establish a system of aeronomic input parameters (i.e., the model of the neutral atmosphere, the solar radiation flux and the rate constants of the principal processes). At the next stage, this system of parameters is substituted in the ionospheric model in order to determine n_c^m and h_m, and the results of the calculations are compared with another set of experimental data, thus confirming the validity of the model which has been developed. This rather complicated method is employed, since, as a rule, direct calculations of the ionosphere not based on a set of self-consistent aeronomic parameters yield erroneous results.

Stubbe /277/, who started out from ionospheric data on n_c^m and h_m, also proposed a method for the calculation of the neutral composition of the upper atmosphere. The validity of his thermospheric model was confirmed by the calculation of the median values of n_c and h_m at a number of stations at moderate and high latitudes. The thermospheric model derived in this way was more realistic, being based on a large number of parameters, than the JACCHIA–71 model which was considered to be the best model until that time, but the calculated values of ionospheric parameters were still unsatisfactory.

Thus, the calculated and the observed values of n_c^m differ by a factor of 2–3 in several cases, while the calculated values of h_m are 50–100 km larger than the values actually observed. According to Stubbe, these discrepancies are mainly due to inaccurate data for the velocities of thermospheric winds, which are calculated according to Jacchia's 1971 model. Stubbe noted in this context that ionospheric data should not only be used to reproduce the neutral composition, but also to obtain the values of the pressure gradients, in order to achieve a correct computation of the thermospheric wind system. However, there may also be other reasons for these discrepancies. Since the significant factors include not only the neutral composition and the winds, but also the ionizing solar radiation which varies from day to day, satisfactory results cannot be expected if this factor is not taken into consideration. Furthermore, the "reconstitution" of the neutral composition involved h_m-data obtained by ionospheric sounding, which are known to be rather unreliable.

Despite the several unsatisfactory results obtained in calculations according to Stubbe's model, the merit of his work lies in his approach to the development of self-consistent models of the ionosphere including a specific example of the construction of such a model. As our knowledge of the processes taking place in the upper atmosphere improves, and the values of aeronomic parameters such as the flux and the spectrum of short-wave solar radiation, thermospheric winds, electric fields and rate constants of ion-molecule reactions become more reliable, theoretical models of the ionosphere will keep on improving, and their utilization in prediction will enable better and more complete ionospheric prediction.

1.3 SHORT-TERM AND LONG-TERM PREDICTION

According to the accepted definition, short-term prediction means monthly, 5-day and half-day predictions. If a longer period of time is involved, it is considered to be long-term prediction. As a rule, the ionospheric parameter which is predicted is the critical frequency of the $F2$-layer, which undergoes considerable irregular variation. The prediction of the critical frequencies of the E- and $F1$-layers merely involves certain semi-empirical calculations, reflecting the connection between the zenith angle of the Sun and the level of the solar sunspot number W /73/. There are many techniques of ionospheric prediction; those employed in the Soviet Union are described in /14, 68/. In the discussion that follows we shall merely note the salient characteristics of these methods.

The short-term predictions now issued are quasi-inertial, i.e., are of the "what-is-to-day-will-be-tomorrow" type, but recurrent magnetic perturbations are allowed for. This applies to all short-term predictions, with the only difference that the montly prediction is recurrence-oriented, while the half-day prediction is essentially quasi-inertial. In the former case the magnetically perturbed days of the coming months are predicted, and the expected deviations of the critical frequency of the $F2$-layer from its median value on these days are specified*. Thus, if negative (or positive) perturbations were observed during the preceding rotations of the Sun, it is very probable that they will be repeated during the rotation period which follows.

The five-day prediction yields the median values of the critical frequency of the $F2$-layer, plotted at hourly intervals during the preceding 10 days. The "moving median"** is computed every 5 days. The data based on the predicted perturbations yield the $\delta f_o F2$ deviations (in marking points).

The half-day prediction yields the estimated state of the ionosphere during the past 12 hours, and the predicted deviations of $f_o F2$ during the coming half-day. If the magnetic field is quiet, while the deviations from the median are the same as on the preceding day, the prediction will be the same as that made for the preceding day.

We shall now examine the assumptions which underlie the long-term prediction of

---

* Median values are obtained from a set of observed values by successively discarding the lowest and the highest values of the parameter, in alternation.

** The "moving median" is the average of a number of observations made over several days. This series, the length of which remains as before, is shifted one day in time to obtain the average for the subsequent day.

median critical frequencies f_o of different ionospheric layers, which is now made 6 months ahead. Thus, the present day MUF prediction /73/ may be regarded as a statistical model, since it constitutes the result of statistical processing of many median values of critical frequencies of the $F2$-layer obtained from the global network of ionospheric sounding stations. The data were broken down by even hours of the local time, 12 months and 4 levels of solar activity. The results of processing of the planetary distribution are given as a set of expansion coefficients by spherical harmonic analysis (Legendre functions). Comparison between the predicted values of median critical frequencies for given levels of solar activity with the results of the observations made at several ionospheric sounding stations showed that the accuracy of the predicted f_oF2 values is fairly high – about 11% /7/. It follows that if the level of solar activity (Wolf number) is accurately predicted, the MUF prediction may yield reliable values of critical frequencies of the $F2$-layer. Another achievement of the MUF prediction is the fact that it yields charts of planetary distribution of MUF's.

We shall now examine the main difficulties and failings of the present-day methods of establishing ionospheric prediction. Irrespective of the particular method employed, there is the problem of predicting the levels of solar and geomagnetic activities. Thus, the sunspot number, W, is an input parameter in the derivation of long-term predictions. On the other hand, the correlation between f_oF2 and W is valid only on the average, and there is no correlation if, for instance, the comparison is made on specific days. The reason for this is that the $F2$-region is sensitive to the flux of solar short-wave radiation, which correlates with the sunspot number only on the average.

In addition, long-term forecasts of W may yield large deviations from the values actually observed. This means that a half-year prediction must subsequently be corrected in accordance with the actual values of W. Furthermore, practical radiocommunication must be based on the true state of the ionosphere on a specific day, which may be very different from the predicted montly average. In practical short-term prediction, based on the moving-median f_oF2, changes in the solar activity level are ignored and, as a result, there are major deviations of the observed f_oF2 values from the median (cf. para. 5.3).

In fact, the daily electron concentrations in the $F2$ layer maximum may differ by more than a factor of two, even on two days in near succession, and even if both days are magnetically quiet. This may be due to changes in the flux of solar radiation, which clearly cannot be taken into consideration when establishing the prediction. It is imperative, therefore, to improve the methods of long-term prediction.

Let us now consider short-term prediction. Apart from predicting perturbations, which is still a major problem, we shall consider the drawbacks of the inertial method of prediction. Thus, the median serving as the reference point from which the deviations δf_oF2 are counted, reflects the perturbed days during the previous 10-day period for which the median was constructed. The result is that if, for example, the following 5 days are quiet, the prediction will be incorrect with respect to these days. Moreover, if more than 5 out of the 10 days which follow prove to be perturbed, the previous median is taken. Clearly, the prediction becomes much less accurate in such cases. This is particularly evident around the equinoxes, when the changes taking place in the upper atmosphere within a few days may be quite considerable.

To sum up this discussion of ionospheric prediction, we shall point out the principal difficulties and the possible ways of improving the existing methods. At present, the ionospheric service furnishes predicted median values of the critical frequency of the layer, as well as its deviations from the median (in point marks). The use of the median is inconvenient in describing the fast-changing conditions in the ionosphere, especially so around the equinoxes, and also under conditions of magnetic perturbations and variations of the solar flux. The operative information does not include the altitude and the half-width data of the $F2$-layer, even though both methods for the computation of these parameters and charts of their global distribution are now available /4/. It is clear that without a knowledge of the distribution of electron concentration with altitude (the $n_e(h)$-profile), an accurate calculation of the propagation paths of radio waves is impossible. Therefore, operative information should include data on the entire profile $n_e(h)$ or at least on a number of its characteristic points such as the maxima of the E, $F1$ and $F2$-layers, the valley and the inflection point in the $n_e(h)$ profile below the $F2$-layer maximum.

Another very complex problem, which is not connected with the calculation and presentation of electron concentration profiles, is the prediction of the solar radiation and the level of geomagnetic activity, the input parameters of several ionospheric models. As already mentioned, variations of the flux of short-wave solar radiaiton led to a number of the variations encountered in the ionosphere. However, this radiation cannot be systematically followed, and its direct prediction a long period of time ahead is difficult (Chapter 3).

The prediction of perturbed conditions in the ionosphere is a problem to which no answer has as yet been found. This is due both to the difficulty of predicting the geomagnetic indexes, A_p and K_p, and to the complexity of planetary distribution and temporal variation of the disturbed conditions in the ionosphere. The onset times of recurrent perturbations can be predicted more less accurately, but only in those branches of the solar cycle where they are recurrent, and this is allowed for in establishing monthly ionospheric predictions. Thus, expected deviations from the median values of the critical frequencies are reported for the days on which the disturbances occur. Practically no other types of ionospheric perturbations may be predicted. Since the ability to predict whether or not an ionospheric disturbance will in fact develop following a recorded flare, depends entirely upon the intuiton and the practical experience of the person making the prediction.

The deterministic method of predicting the state of the ionosphere (Chapter 5) is based on the same principle as the present-day long-term prediction of the median, f_oF2; the indexes of the short-wave and corpuscular solar radiations, rather than the radiations themselves, are chosen as the input parameters for this purpose. The sunspot number, W, has been replaced by the $F_{10.7}$ flux of radio wavelength radiation; unlike W, the parameter $F_{10.7}$ reflects the variations in the short wave flux not only on the average, but on individual days. Thus, under similar helio-geophysical conditions it is possible not only to derive the monthly average state of the ionosphere, but also to use the daily changes of $F_{10.7}$ (under otherwise similar conditions) to predict changes in f_oF2 and other ionospheric parameters.

The values of $F_{10.7}$ and A_p may be fairly reliably predicted one or two days ahead, so that a deterministic calculation of ionospheric parameters under given helio-geophysical conditions can only effectively be made 1–2 days ahead.

CHAPTER 2. THE THEORY OF IONOSPHERE FORMATION

Early studies of the ionosphere were based on data obtained by radio-sounding from the ground, i.e., mainly on the values of the critical frequency of the main layers of the ionosphere. The ideas about the nature of the processes taking place at ionospheric altitudes were incomplete, and could not always explain the complex and variegated behavior of the ionosphere.

Thus, these early ideas about the mechanism of ionosphere formation were substantially modified according to direct measurements of short-wave solar radiation and altitude distribution of the ion and neutral composition in the upper atmosphere, made with the aid of rocket-borne instruments. Since the principal mechanisms of formation of the ionospheric layers, E, $F1$, $F2$ and the topside ionosphere situated above the maximum of the $F2$-layer are now reasonably clear, and a basic outline of the general theory of ionosphere formation is now available, present research efforts are focused in the direction of concrete applications of general theoretical assumptions, and explanation of known variations of the ionosphere as a function of the helio-geophysical conditions. In this chapter we present the fundamental principles of the general theory of ionosphere formation, required to understand the role played by specific processes causing the observed variations of ionospheric parameters. Reference will be made to special monographs in which the problems discussed in this book are treated in greater detail.

The main ionospheric layers occupy a region between 100 and 600 km in altitude. As we move from the lower to the upper boundary of this region, the concentration of neutral particles n, the composition and the temperature of the atmosphere undergo considerable changes. Thus, below $n \approx 10^{12}$ cm^{-3} the principal species are molecular N_2 and O_2 and the temperature is about 300°C, while at the upper boundary of the region where n $\approx 10^8 - 10^6$ cm^{-3}, the atmosphere consists mainly of atomic oxygen, nitrogen, helium and hydrogen, and the temperature varies between 750 and 1500°K, depending on the level of solar activity. Furthermore, the parameters of the upper atmosphere vary from day to day with the helio-geophysical conditions, and these changes are reflected by the state of the ionosphere.

2.1 THE PRINCIPAL PROCESSES

The theory of ionosphere formation is mainly based on the so-called continuity or balance equation, which describes the mechanism of variation of a given ionized component with respect to three principal processes – formation, transport and recombination of ions and electrons. If V_i is the bulk transport velocity, the concentration change produced by this process equals the divergence of the flux, $n_i V_i$. Using the standard notation, q and L, for the rates of formation and recombination of ions and electrons respectively, the continuity equation may be written as follows:

$$\frac{\partial n_i}{\partial t} = q_i - L_i - \text{div}(n_i V_i), \tag{2.1}$$

where n_i is the ion concentration.

Equations such as (2.1) are usually written for the i-th ionized species, under conditions of charge neutrality of the plasma, $n_e = \Sigma n_i$.

The various ionospheric regions are formed by the effect of various parts of the solar UV spectrum, and have different ion compositions. Their formation mechanisms are also not the same. Depending on the altitude interval considered, only certain terms in equation (2.1) may be significant. Thus, for instance, in E and $F1$-regions it is permissible to neglect the transport processes as compared with the other two processes, while the typical recombination time constants for the principal ions NO^+ and O_2^+, are so short that, under normal conditions, $\partial n_i/\partial t = 0$, and the equilibrium concentrations are determined by the condition of photochemical equilibrium, $q_i = L_i$. On the other hand, in the topside ionosphere transport is the principal process, while the processes of ionization and recombination are insignificant. However, in the $F2$-layer maximum all these three processes must be considered simultaneously. Effects related to the non-stationary state are also significant, so that all the terms of equation (2.1) must be used. We shall begin by considering the photochemical processes.

2.1.1. Processes of Ionization and Recombination. In daytime it is the short-wave solar radiation which acts as the main source of ions at altitudes of $h > 100$ km. At night the E-region displays significant ionization produced by corpuscular and scattered UV radiations. In the course of the ionization of neutral atmospheric components the primary ions, O^+, O_2^+ and N_2^+, are formed. Following ion-molecular reactions they are ultimately converted to NO^+ ions, which are then lost in a dissociative recombination reaction /33, 50, 65/. Some of the primary molecular ions O_2^+ and N_2^+ are directly lost in the dissociative recombination proces.

The formation of primary ions is induced by $\lambda < 103.76$ nm radiation /33/. The λ-values corresponding to ionization thresholds for the principal neutral atmospheric components are:

Component	O_2	N_2	O	N	NO
λ_i, nm	102.7	79.6	91.2	85.2	134.0

A process which is significant in the lower ionosphere is the ionization by the Lyman α-line ($\lambda = 121.6$ nm) of a very minor atmospheric component, NO. The L_α-line is the strongest one in the spectrum of short-wave solar radiation.

On passing through an atmospheric layer dh, with a particle concentration of n cm^{-3}, radiation of intensity I_λ becomes attenuated by the magnitude

$$dI_\lambda = -\sum_j \sigma_{aj\lambda} I_\lambda n_j dh = -I_\lambda d\tau_\lambda, \tag{2.2}$$

where τ_λ is the optical depth of the atmosphere at the altitude h and $\sigma_{aj\lambda}$ is the absorption cross-section.

In order to describe the different conditions of solar illumination of the upper atmosphere, the expression (2.2) is written as:

$$dI_\lambda = -I_\lambda \mathrm{Ch}(\chi) d_\tau, \qquad (2.3)$$

where the dimensionless magnitude $\mathrm{Ch}(\chi)$ is a measure of the variation of the depth of the atmosphere with the solar zenith angle, χ, known as the Chapman function. If $\chi \leqslant 70°$, the Chapman function is approximately equal to $\sec \chi$. However, around sunrise and sunset the values of $\mathrm{Ch}(\chi)$ are very different from $\sec \chi$. If the zenith angle $\chi > 70°$, the following approximation is employed /65/

$$\mathrm{Ch}(\chi) \approx \sqrt{\frac{\pi x \sin \chi}{2}}\, e^{1/2\, x \cos^2 \chi}\, [1 \pm \mathrm{erf}(\tfrac{1}{2} x \cos^2 \chi)^{1/2}] \qquad (2.4)$$

where $x = (R_0 + h)/H$; R_0 is the radius of the Earth; and H is the atmosphere scale height. The signs, $\pm$, correspond to cases where $\chi \gtrless 90°$, respectively.

The production of the j-th ionic species in $1\ \mathrm{cm}^3$ per unit time by the ionizing radiation with a wavelength λ is:

$$q_{j\lambda} = \sigma_{ij\lambda} I_\lambda n_j$$

where $\sigma_{ij\lambda}$ is the ionization cross-section. If I_λ, obtained by integrating (2.3), is substituted into this expression, the rate of ion production by monochromatic radiation of wavelength λ may be written as:

$$q_{j\lambda}(h) = n_j \sigma_{ij\lambda} I_{0\lambda} e^{-\tau\lambda}$$

If we consider the entire spectrum, from the threshold, λ_i, up to X-rays with the wavelength, λ_r, and bear in mind that the neutral atmosphere is a multi-component system, we obtain the final expression for the rate of production of ions of the j-th speicies, which is commonly employed in calculations for pratical purposes:

$$q_j(h) = n_j \sum_{\lambda_{ij}}^{\lambda_r} I_{0\lambda} \sigma_{ij\lambda} \exp \left[-\sum_j \sigma_{aj\lambda} \int_h^\infty \mathrm{Ch}(\chi) n_j dh \right] \qquad (2.5)$$

For a more detailed review of the ionization of atmospheric components both by the solar and by the corpuscular radiation see /33, 65/. Table 2.1, which is taken from /25/, shows data on the absorption cross-sections and ionization cross-sections of the principal atmospheric components as a function of the spectral interval. The radiation intensities, I_λ, for the spectral region at 4 levels of solar activity $F_{10.7}$ /36/* are also shown.

Figure 2.1 shows the overall ion production rates $(q(O^+) + q(O_2^+) + q(N_2^+))$ for high $(F_{10.7} = 200)$ and low $(F_{10.7} = 80)$ levels of solar activity, computed for a number of solar zenith angles. The computations are based on the thermospheric MSIS model /177/

* According to the most recent data (cf. para. 3.1) the radiation intensity I_λ should be about twice as large. The flux $F_{10.7}$, here and the text that follows is expressed in units of 10^{-22} W/(Hz·m²).

during an equinoctial period, corrected for diurnal variations of composition and temperature.

It is seen from Fig. 2.1 that in the zone of the $F2$-layer, i.e., above the maximum rate of ion production, lg q decreases almost linearly with h. This is due to the fact that at such altitudes the atmosphere is almost isothermal, with atomic oxygen as its principal component, while the absorption of the radiation is negligible. The important consequence of this fact is that in calculating the rate of ion production in the $F2$-region only the value of the total flux of solar radiation is required – the energy distribution over the spectrum is largely immaterial. Consequently, in numerical predictions of the state of the $F2$-region, reducing the number of spectral intervals of the ion-production function saves a great deal of time.

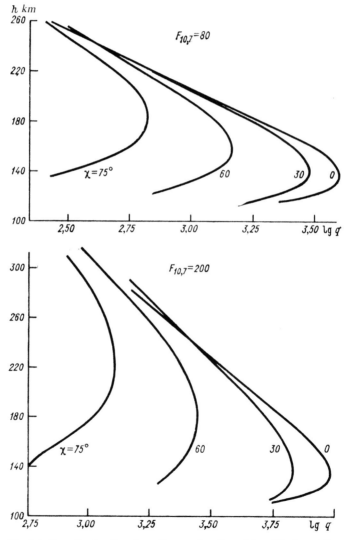

Fig. 2.1. Overall rate of ion formation as a function of the altitude and the zenith angle for the equinox period, at various levels of solar activity.

TABLE 2.1

Absorption cross-sections σ_a and ionization cross-sections, σ_i of the major atmospheric components, and intensity of short-wave solar radiation, I_λ, for different $F_{10.7}$-values.

λ nm	$\sigma_a, 10^{19}\,cm^2$			$\sigma_i, 10^{19}\,cm^2$			$I_\lambda, 10^{-7}\,WB/(sec \cdot cm^2)$			
	O	O_2	N_2	O	O_2	N_2	$F_{10.7}=100$	$F_{10.7}=120$	$F_{10.7}=144$	$F_{10.7}=200$
103.76		7.8	0		0.008		113	140	170	206
103.19		10.4	0		0.10		153	190	230	280
102.57		15	0		9.7		268	310	350	396
102.6–100.5		11.2	0		7.0		31.5	36.5	41.5	46.7
100.5–100.3		75	0		21		1.9	2.2	2.5	2.8
100.3–99.3										
99.15		18.5	0.75		12		39	45.9	53	61.2
99.30–99.14		100	40		20		0	0	0	0
99.14–98.67		37.2	1.5		4.1		6	7	8	9
98.5–98.4		37.2	30		4.1		0	0	0	0
98.67–98.5		14	1.67		9.5		13.8	16.4	19	22
98.98		185	1		90		0	0	0	0
98.4–98.2		30	350		20		3.8	4.4	5	5.7
98.2–97.83										
97.83–97.65										
97.7		40	0.82		24.7		321	379	441	514
97.60–97.25		112	1.0		2.0		0	0	0	0
97.50–97.25		220	350		100		0	0	0	0
97.25		320	$3 \cdot 10^3$		200		61.2	70.8	80	90.5
97.25–97.07		46	150		36.1		0	0	0	0
97.07–96.8		26	1.85		26		2.3	2.6	3	3.4
96.8–96.5		220	380		80		2.3	2.6	3	3.4
96.5–96.3		220	1.6		80		1.5	1.8	2	2.3
96.3–95.95		150	250		30		3	3.6	4	4.6
95.95–95.75		200	120		170		1.5	1.8	2	2.3
95.75–95.6		445	0.75		380		0	0	0	0
95.6–95.45		200	28		170		1.1	1.3	1.5	1.7
95.45–95.05		33.5	0.55		27		3	3.6	4	4.6
95.05–94.9		33.5	15		15		0	0	0	0
94.97		63	52		30		29.8	34.4	39	44.2
94.9–94.85										
94.75–94.70		185	0.75		100		0	0	0	0

Range									
94.85–94.75		370	0.75		200	0	0	0	0
94.7–94.55		200	13		180	0	0	0	0
94.55–94.4		37.2	2.3		37	6.9	8.4	10	11.9
94.4–94.18		33.6	95		31	0	0	0	0
94.18–94.02		63	38		32	0	0	0	0
94.02–93.75		200	400		180	16.8	19.4	22	25
93.78		50	100		29	0	0	0	0
93.75–93.55		37.2	0.50		27	9	11	13	15.5
93.55–93.45		45.5	18.5		27	9.9	11.5	13	14.7
93.45–93.32		112	0.55		50	0	0	0	0
93.32–93.05		150	37		90	9.9	11.5	13	14.7
93.05–92.83		224	270		140	0	0	0	0
92.83–92.65		483	2.6		34	2.1	2.5	3	3.5
92.65–92.6	0	100	34	0	50	0	0	0	0
92.6–92.4	0	112	6.5	0	60	28.3	32.7	37	42
92.4–92.23	0	100	83	0	50	3.8	4.4	5	5.7
92.23–91.96		43	15		25	0	0	0	0
91.96–91.9		65	4.0		28	13.9	16	18.2	20.6
92.225–92.19	0	100	78	0	50	109.4	126.6	143.5	162.7
91.9–91.6		43	1.9		36	40.5	46.8	53	60.1
91.0–91.2	0	80	250	0	37	71	82	93	105
91.6–91.4	0	155	8.5	27	114	13.5	15.6	17.7	20
91.4–91.2		93	41	27	56	11.2	12.9	14.6	16.5
91.0–90.9	27	63	11	27	34				
90.9–90.7		75	95		45				
90.6–90.0	27	110	8.5	29	37				
90.7–90.6	27	82	48	28	70				
89.08–88.78	29			28					
90.0–89.55	28								
86.86–86.5	28								
89.55–89.42									
89.42–89.31									

λ nm	σ_a, 10^19 cm^2			σ_i, 10^19 cm^2			I_λ 10^-7 KW/(sec·cm^2)			
	O	O_2	N_2	O	O_2	N_2	$F_{10.7}=100$	$F_{10.7}=120$	$F_{10.7}=144$	$F_{10.7}=200$
89.31–89.21	29	82	9.3	29	55		39.4	45.5	51.6	58.5
88.56–88.33										
87.05–86.86										
89.21–89.08	29	82	52	28	70		12.6	14.5	16.5	18.7
88.78–88.56	28.5	170	93	28.5	100		19.3	22.2	25.2	28.5
88.33–88.06	29	67	110	29	33		21.2	24.5	27.8	31.5
88.06–87.77	29	130	10	29	90		22	25.4	28.8	32.6
87.77–87.58	29	86	180	29	50		29.4	34	38.5	43.6
87.29–87.05										
87.58–87.29	29.5	75	5.6	29.5	38		20	23	26	29.4
86.5–85.83	30.5	82	6.5	30.5	33		35.9	41.5	47	53.2
85.83–85.31	31	105	110	31	37		34.2	39.6	45	51.1
85.31–85.1	31.5	160	18	31.5	39		13	15	17	19.3
84.38–84.28										
85.1–84.95	31	140	165	31	42		12.6	14.8	17	19.5
84.95–84.48	31.5	120	47	31.5	41		18.7	21.6	24.5	27.8
84.48–84.38										
77.0–76.75										
75.9–75.67	34	220	140	34	100		10.6	12.2	13.9	15.8
75.1–74.55										
84.28–84.03	32	300	130	32	35		9	10.4	11.8	13.4
84.03–83.86	32	170	14	32	120		5.8	6.7	7.6	8.6
83.86–83.7	32	185	82	32	36		5.3	6	6.9	7.8
83.7–83.56	32	150	9.3	32	44		4.5	5.2	5.9	6.7
83.56–83.46	32	185	110	32	43		12.4	14.6	17	19.6
83.46–82.75	32		380	32	60		49.2	57.5	66	75.7
82.75–82.4										
78.3–78.18	32	220	150	32	85		11.5	13.2	15	17
76.0–75.9										
75.67–75.57										
82.4–82.13	32	260	3.7	32	74		6.6	7.6	8.7	9.8

82.13–81.82	32.3	300	130	32.3	70		6.9	8	9.1	10.3
81.82–81.42	32.5	330	7.8	32.5	110		8.2	9.5	10.8	12.2
81.42–81.25	32.5	330	300	32.5	100		3.2	3.7	4.2	4.8
81.25–80.75	32.5	410	15	32.5	120		9.2	10.6	12	13.6
80.75–80.55	33	520	130	33	70		3.4	3.9	4.4	5
80.55–80.23	33	410	56	33	85		5.1	5.9	6.7	7.6
80.23–80.05	33	370	340	33	100		2.7	3.2	3.6	4.1
80.05–79.53	33	330	47	33	100		7.1	8.2	9.3	10.6
79.53–78.9	33	305	280	33	120	100	26.3	31.2	36.4	42.5
78.9–78.47	33.5	285	120	33.5	110	75	19.5	23.1	27	32.5
78.47–78.3	33.5	280	670	33.5	120	400	2.1	2.5	2.8	3.2
78.18–78.07										
77.04										
76.6–76.31		230	300			250	3.9	4.5	5.1	5.8
75.57–75.4		280	140		92	80	11.1	13.6	16.5	19.8
78.07–77.7	33.5	185	450	33.5	100	280	3.5	4.2	4.8	5.6
77.7–77.44	34	185	100	34	92	72	1	1.1	1.3	1.5
77.44–77.3	34	275	450	34	105	200	1.53	1.76	2.06	2.27
77.3–77.1	34	275	100	34	85	78	0.76	0.9	1	1.13
77.1–77.0	34	220	115	34	85	100	1	1.19	1.35	1.53
76.75–76.6	34	240	130	34	67	78	15	19.2	23.6	29.1
76.31–76.23	34	230	230	34	78	110	6.8	8.3	9.8	11.6
76.23–76.0	34.5	205	185	34.5	80	110	1.4	1.6	1.8	2
75.4–75.1	34.5	260	370	34.5	110	300	0.9	1	1.15	1.3
74.55–74.32	35	260	240	35	130	200	1.5	1.8	2.1	2.4
74.32–73.9	35	410	250	35	250	220	9.15	11	13	15.3
73.9–73.2	75	370	300	75	200	220	1.9	2.2	2.5	2.8
73.2–72.5	77	330	560	77	155	430	0.9	1	1.2	1.4
72.5–72.1	82	410	480	82	200	400	2	2.4	2.7	3
69.58–69.38	83	247	380	83	150	310	1.3	1.5	1.7	1.9
68.26–68.1	83.5	253	335	83.5	170	265	1.3	1.5	1.7	1.9
67.57–67.47	84	270	315	84	170	255	0.5	0.58	0.7	0.75
67.18–67.08	82	250	260	82	200	230	15.2	18.1	21	24.4
72.1–66.5	120	380	335	120	190	280	127	160.1	202	237
66.5–60.0	130	230	230	130	230	230	67	78	89	102
58.43										
60.0–58.0	129	270	280	129	260	280	7.4	10.3	13	16.7

λ hm	$\sigma_a, 10^{19} \mathrm{cm}^2$			$\sigma_i, 10^{19} \mathrm{cm}^2$			$I_\lambda, 10^{-7}\,\mathrm{KW/(sec \cdot cm^2)}$			
	O	O_2	N_2	O	O_2	N_2	$F_{10.7}=100$	$F_{10.7}=120$	$F_{10.7}=144$	$F_{10.7}=200$
58.0–54.0	130	300	260	130	260	260	28.6	34.6	41	48.7
54.0–51.0	130	290	230	130	260	230	12.4	19.2	27	50.1
51.0–50.0	130	280	200	130	250	200	16.8	19.7	22.6	25.9
50.0–48.0	129	270	170	129	240	170	40.9	58.6	79	112
48.0–46.0	121	260	160	121	230	160	13.8	17.2	21	25.7
46.0–43.5	105	250	160	105	230	160	18.8	23.2	28	33.7
43.5–40.0	125	240	150	125	240	150	10.3	16.2	23	34.1
40.0–37.0	111	230	150	111	230	150	7.6	9.7	12	14.8
36.81	103	220	113	103	220	113	34.6	45.2	56	69.4
37.0–35.5	100	220	140	100	220	140	64	89.3	118	162.4
35.5–34.0	93	220	110	93	220	110	42.6	56.4	70	86.8
34.0–32.5	87	210	100	87	210	100	24.4	46.2	72	116.6
32.5–31.0	81	200	85	81	200	85	31	40.3	50	62.0
30.38	98	166	50	98	166	50	389	463	540	629
31.0–28.0	92	187	65	92	187	65	53	91	135	209
28.0–26.0	80	160	60	80	160	60	44.5	59.5	72.5	95.4
76.51	34	230	800	34	123	660	12.7	15.2	18	21.3
25.63–25.7	72	144	60	109	220	91	27.5	38	49	65.5
26.0–24.0	67	134	60	104	208	93	52	72	97	134.7
24.0–22.0	56	112	55	94	188	93	52	69.6	87	110.7
22.0–20.5	47	94	55	85	170	100	41.3	56.2	74	99.2

	40	80	50	79	158	100	62.5	85.5	112	150
20.5–19.0	40	80	50	79	158	100	62.5	85.5	112	150
19.0–18.0	34	68	40	71	142	84	92.5	120	150	188.5
18.0–16.5	29	57	40	65	130	90	78.4	102	129	164
16.5–13.8	26	52	40	67	134	103	15.6	21.5	28.5	39.3
13.8–12.0	19.5	39	27	58	116	80.3	5	6.9	9.1	11
12.0–9.2	11.5	23	14	43.2	86.3	52.6	9.6	12.3	15	18.3
9.2–7.1	6.75	13.5	8.1	31.5	63	37.8	12.8	16.9	21	25.8
7.1–5.6	3.3	6.6	4.0	20.6	41.1	24.1	6.8	9.6	12.6	16.6
5.6–4.2	1.75	3.5	2.15	14	28	17.1	8	10	12.3	16.8
4.2–3.1	0.875	1.75	1.0	9.4	18.8	10.7	3.6	4.7	6.1	8.7
3.1–2.6	0.485	0.97	12	6.7	13.4	166	0.9	1.3	1.6	2.7
2.6–2.28	0.34	0.68	8.7	5.25	10.5	135	0.34	0.54	0.70	1.3
2.28–1.75	4.65	9.3	5.5	88	176	104	0.33	0.53	0.70	1.4
1.75–1.45	2.15	4.3	2.75	52	104	66	0.081	0.133	0.188	0.41
1.45–1.15	1.25	2.5	1.6	37.1	74.2	47.6	0.036	0.061	0.091	0.21
1.15–0.9	0.65	1.3	0.83	25	50	32.1	0.012	0.022	0.035	0.086
0.9–0.7	0.35	0.70	0.46	17	34	22.3	0.0025	0.0054	0.0098	0.03
0.7–0.55	0.17	0.34	0.20	11	22	13	0.0005	0.0011	0.0025	0.01
0.55–0.45	0.09	0.187	0.105	7.0	14	8.1	0.0001	0.00018	0.00045	0.0022

Mass-spectrometric measurements taken on board rockets show that the ion concentration in the ionosphere is not proportional to the rates of production of the principal ions, and that the concentration NO^+ ions is high, even though neutral nitrogen oxide is only a minor atmospheric component (its relative concentration is $10^{-3} - 10^{-4}$ even at the altitude of the E-layer). This proves that the ion-molecule reactions taking place in the ionosphere result in effective transformations of one ionic species into another.

Theoretical studies indicate that the principal ion-molecule reactions determining the interaction between the primary ions, O^+, N_2^+ and O_2^+ and the neutral atmospheric components are:

$$O^+ + N_2 \rightarrow NO^+ + N, \gamma =$$

$$= \begin{cases} 1.2 \cdot 10^{-12}(T_{\text{eff}}/300)^{-1.0}, \ T_{\text{eff}} < 740 \text{ K} \\ 8.0 \cdot 10^{-14}(T_{\text{eff}}/200)^{2.0}, \ T_{\text{eff}} > 740 \text{ K} \end{cases} /215/ \tag{2.6}$$

$$O^+ + O_2 \rightarrow O_2^+ + O, \gamma_2 = 1 \cdot 10^{-9}/T_n^{0.7} \ /155/; \tag{2.7}$$

$$N_2^+ + O \rightarrow NO^+ + N, \gamma_3 = 1.4 \cdot 10^{-10} \ /152/ \tag{2.8}$$

$$N_2^+ + O_2 \rightarrow O_2^+ + N_2, \gamma_4 = 6.0 \cdot 10^{-11}(T/300)^{0.6} \ /195/ \tag{2.9}$$

$$O_2^+ + NO \rightarrow NO^+ + O^2, \gamma_5, = 6.3 \cdot 10^{-10} \ /153/ \tag{2.10}$$

$$O_2^+ + N \rightarrow NO^+ + O, \gamma_6 = 1.8 \cdot 10^{-10} \ /158/. \tag{2.11}$$

The γ-value for each of the above reactions is shown. This constant determines the reaction rate in 1 cm^3 during one second, i.e., the rate of conversion of the reacting primary ions, which equals the rate of production of the ions obtained during the course of the reaction. Data yielded by observations on the concentrations of charged and neutral particles prove that at altitudes above 120–140 km the main reactions involved are the first three reactions, whereas at lower altitudes the last three reactions are also significant.

In order to decide the relative importance of any specific reaction, reliable data on the concentrations of neutral particles and ions, as well as the values of the rate constants of ion-molecule reactions must be available. The greatest source of uncertainty is the value of γ_1 in the first reaction which, together with the second reaction, are the main routes by which O^+ ions are lost in the $F2$-region of the ionosphere. In the past there were no reliable laboratory measurements of γ_1, and for this reason research work was directed at determinations of the different values of γ_1 and its variation with the temperature (cf. Chapter 3).

The final product of the reactions discussed above is the NO^+ ion. This ion is not converted into other ions, and since it has the lowest ionization potential, it is lost by dissociative recombination. Some of the O_2^+ ions, both the primary ions and those formed by ion-molecule reactions, are also lost in the same manner. The primary N_2^+ ions are lost at altitudes below 200–250 km principally by reactions (2.8) and (2.9), while at higher altitudes the main mechanism is dissociative recombination.

Thus, in the altitude interval under study, three processes of dissociative recombination constitute the final stage of ionic transformations /100/:

$$
\left.
\begin{aligned}
O_2^+ + n_e &\rightarrow O + O & \alpha_1 &= 2.2 \cdot 10^{-7}\,(T_e/300)^{-0.7} \\
NO^+ + n_e &\rightarrow N + O & \alpha_2 &= 4.5 \cdot 10^{-7}(T_e/300)^{-0.83} \\
N_2^+ + n_e &\rightarrow N + N & \alpha_3 &= 3.0 \cdot 10^{-7}\,(T_e/300)^{-1.5}
\end{aligned}
\right\} \quad (2.12)
$$

The rate constants, α_1, α_2 and α_3, depend on the electron temperature, T_e; consequently, since T_e varies in the course of the day, so do the constants. It is especially important to bear this in mind in the $F1$ region, where the electron temperature undergoes significant diurnal variations.

Under equilibrium conditions in the $E1$ and $F1$-regions, the rates of ion formation equal the rates of ion loss, and then the following expressions for the concentrations of the principal ions are obtained:

$$[O^+] = q(O^+)/(\gamma_1[N_2] + \gamma_2[O_2]), \qquad (2.13)$$

$$[N_2^+] = q(N_2^+)/\gamma_3[O] + \gamma_4[O_2] + \alpha_3 n_e) \qquad (2.14)$$

$$[O_2^+] = (q(O_2^+) + \gamma_4[N_2^+][O_2] + \gamma_2[O^+][O_2])/(\mu + \alpha_1 n_e) \qquad (2.15)$$

$$[NO^+] = (\gamma_1[O^+][N_2] + \gamma_3[N_2^+][O] + \mu[O_2^+])/\alpha_2 n_e \qquad (2.16)$$

Here, $\mu = \gamma_5[NO] + \gamma_6[N]$ is the effective rate of transformation of O_2^+ into NO^+ ions. This magnitude must be introduced, since the concentrations of the minor atmospheric components, NO and N, is not known with sufficient accuracy, while the role they play in ion transformations is significant. In order to be able to estimate the contributions of reactions (2.10) and (2.11) to the overall transformation rate of O_2^+ ions, values of μ are determined from the experimental values of the ion composition, as suggested in /21/.

It may be seen from equations (2.13) to (2.16) that the concentration of O^+ ions may be directly determined from the ion production rate and concentrations of neutral particles, whereas the concentrations of O_2^+, NO^+ and N_2^+ may be determined by solving a set of equations.

Using the photochemical theory models of the altitude distributions of ions and electrons in E and $F1$-regions of the ionosphere are readily calculated. For a more detailed analysis of the photochemical theory of the ionosphere and the atmosphere see monographs /11, 50/.

2.1.2. Plasma Transport in the Ionosphere. The charged particles, formed in a definite volume of the ionosphere as a result of ionization may immediately recombine or may be shifted to other regions of the ionosphere by dynamic processes. At the altitudes of the E and $F1$ ionospheric layers where the concentration of neutral particles is sufficiently high, the transport processes are insignificant. Above 150–200 km the characteristic transport time constants are of the same order as the characteristic recombination times, and equilibrium ion concentrations are no longer determined by local processes. It is therefore necessary to take into consideration the transport of plasma from its site of formation to another region, where recombination takes place.

In the mid-latitude ionosphere, which is the subject of this book, plasma is mainly transported vertically. Such transport processes are responsible for the $F2$-region formation. In the $F2$-region plasma is transported by diffusion in the gravitational field, and also as a result of thermospheric winds and electric fields. The vertical component of the bulk velocity of the O^+ ion, the major ion in this region, is given for the unidimensional case /91/ by the expression

$$v_z = -D_{in}\sin^2 I \left[\frac{1}{n_i} \frac{\partial n_i}{\partial z} + \frac{T_e}{n_e T_i} \frac{\partial n_e}{\partial z} + \frac{m_i g}{kT_i} + \right.$$

$$\left. + \frac{1}{T_i} \cdot \frac{\partial(T_e + T_i)}{\partial z} \right] + \frac{E_y}{B} - \cos I + v_{nx} \sin I \cos I \cos D -$$

$$- v_{ny} \sin I \cos I \sin D. \qquad (2.17)$$

This equation is based on the assumption that the x-axis points southwards, the y-axis eastwards, and the z-axis upwards; I and D are the magnetic dip and declination angles of the line of force, respectively, g is the gravity acceleration, m_i is the mass of the O^+-ion, k is the Boltzmann constant and T_e, T_i and T_n are the temperatures of the plasma and neutral gas respectively. The coefficient D_{in} in the diffusion rate equation is defined as /91/:

$$D_{in} = kT_i/\mu_{in}\sum_j v_{ij}$$

In the above equation, the denominator stands for the sum of the collision frequencies of the O^+ ion and the neutral components, O, O_2 and N_2 which, according to /91/, may be written as follows:

$$\mu(O^+, O) = 3.35 \cdot 10^{-11}(T_i + T_n)^{1/2} [O],$$

$$\mu(O^+, O_2) = 1.08 \cdot 10^{-9} [O_2],$$

$$\mu(O^+, N_2) = 0.95 \cdot 10^{-9} [N_2].$$

Thermospheric winds are the second most important diffusional effect, since by their motion they entrain the plasma which, being magnetically dominated, can only move in the $F2$-region along magnetic lines of force, and may thus substantially affect the altitude distribution of electron concentration, as well as the $F2$-layer itself at night time. Thermospheric winds are caused by heating and expansion of the upper atmosphere. The resulting horizontal pressure gradients serve as the motive force for the thermospheric winds. The wind velocity in the horizontal directions can be described /276/ by the following system of equations:

$$\frac{\partial v_{nx}}{\partial t} - 2\Omega v_{ny}\sin\varphi = \frac{\eta}{\rho}\frac{\partial^2 v_{nx}}{\partial z^2} - \frac{1}{\rho}\frac{\partial p}{\partial x} -$$

$$- \frac{1}{\rho}\sum_i (v_{nx} - u_{ix}\sum_j n_i v_{ij}\mu_{ij}$$

$$\frac{\partial v_{ny}}{\partial t} - 2\Omega v_{nx}\sin\varphi = \frac{\eta}{\rho}\frac{\partial^2 v_{ny}}{\partial z^2} - \frac{1}{\rho}\frac{\partial p}{\partial y} - \tag{2.18}$$

$$- \frac{1}{\rho}\sum_i (v_{ny} - u_{iy}\sum_j n_i v_{ij}\mu_{ij}$$

where v_{nx} and v_{ny} are meridional and zonal velocity components, Ω is the angular velocity of the Earth, η is the viscosity, ρ and p are the mass density and the neutral gas pressure, respectively, $Gm_{ij} = m_i m_j/(m_i + m_j)$ is the reduced mass, v_{ij} is the collision frequency of ions with neutral particles, and u_{ix} and u_{iy} are the velocities of the ions along these routes.

It may be seen from equations (2.18) that thermospheric wind velocities are not determined by pressure gradients alone, but also by the viscosity, by the Coriolis force and by the ion drag; if the aforementioned factors are not taken into account, the wind will be geostrophic, i.e., directed along the isobars. As a result of the ion drag there is a wind component perpendicular to the isobars and quantitatively of the same order of magnitude as the geostrophic component. If the charged particle concentrations are higher than $2\cdot10^5$ cm^{-3}, the effect of the ion drag predominates, and the direction of the wind becomes perpendicular to the isobars.

Theoretical calculations /78, 103, 109, 110, 161, 162, 202, 245, 248, 273/ and observations made by the method of incoherent scatter /79, 80, 140, 174, 257, 258/ indicate that the characteristic wind velocities in the mid-latitude $F2$-region are 30–50 m/sec by day and 150–200 m/sec at night. It should be pointed out that planetary circulation and observed wind velocities vary considerably with solar and geomagnetic activities, and also differ according to the seasons of the year /80, 257, 258/. Thus, in summer, when the high-latitude regions are hot almost around the clock, the wind blows towards the equator for the greatest part of the day. A similar situation is noted /15, 254/ for periods of geomagnetic perturbation, when the polar regions undergo intensive heating. Under ordinary conditions, in winter and during equinoctial periods, the wind in middle latitudes blows towards the pole by day and towards the equator at night.

Apparently ion drag has a significant effect on the velocity of the thermospheric wind, and, in turn, the winds themselves affect the electron concentration and its altitude distribution. Consequently, in order to be able to calculate wind velocity and electron concentration correctly, a self-consistent problem should be solved. However, the main difficulty involved in the calculation of thermospheric winds resides in the fact that the present-day models of the thermosphere, while giving a fairly faithful reproduction of neutral composition and temperature, are inaccurate where pressure gradients are concerned. Thus, for example, it has been shown /277/ that if the error in the density values at 350 km altitude is +10% at the equator and −10% at the pole, the resulting error in the pressure gradients will be of the same order of magnitude as the gradient itself. In view of the high standards of accuracy which the thermospheric models have to meet, computation errors are clearly inevitable.

Moreover, the system of thermospheric winds is determined not only by the heating of the upper atmosphere by short-wave solar radiation, but also depends on the Joule heating and ion convection in high-altitude regions /87, 117, 151/, so that the global wind chart may prove to be exceedingly complex.

The interaction between the ionosphere and the protonosphere represents an interesting aspect of the plasma dynamics. The protonosphere is the outer part of the ionosphere, in which light H^+ ions are the major component. Thus, the protonosphere acts as a reservoir, which fills with plasma from the $F2$-region in the sunlight, while the plasma moves back into the ionosphere at night. According to the experimental recordings of "whistles" (atmospherics) /230/, and the results of direct determinations of the velocity of the plasma by the incoherent scatter method /146/, typical values of plasma fluxes between the ionosphere and the protonosphere are about 10^8 cm^{-2}·sec^{-1}.

The effects of ionospheric-protonospheric fluxes on the $F2$-region has been studied by a large number of workers (cf. para. 2.3.2). It was shown, as a result, that under normal daytime conditions fluxes of the order of 10^8 cm^{-2}·sec^{-1} do not have a significant effect on the $F2$-region. Their effects begin to be noticeable as the so-called critical value is approached. For the daytime $F2$ region, this value is about $2·10^9$ cm^{-2}·sec^{-1} /165/. Certain effects of protonospheric fluxes in the $F2$-region will be discussed in para. 2.3.2.

Plasma displacement in the $F2$-region is also caused by electromagnetic drift across electrical and magnetic fields. Data on the electric fields at altitudes corresponding to the $F2$-region are obtained with incoherent scatter instruments /102, 107, 167/. The vertical drift induced by the electric field affects the $F2$-region in a similar manner to the winds. It is thought /142, 146/ that during magnetically quiet periods the electric fields are weak at latitudes (less than about 1 mV/m) and, in the case of the Millstone Hill station, in particular, their contribution to the observed vertical drift rate is less than 10 m/sec.

2.1.3. The Temperature of Electrons and Ions. In the early 1960's, advances in the measurement techniques at ionospheric altitudes allowed to observe that the different components of the upper atmosphere are at different temperatures. It was found that above an altitude of about 150 km, the electron temperature T_e becomes much higher than the temperature of neutral particles T_n and the ion temperature, T_i. With increasing altitude, differences between T_e and T_n, and between T_e and T_i continue to increase, and begin to decrease, respectively.

Figure 2.2 shows the profiles of T_e and T_i, determined by the incoherent scatter method /146/, and the profile of neutral temperature calculated for that day in accordance with the thermospheric MSIS model. It should be pointed out that the difference between those two temperatures also persists at night; moreover, at higher altitudes there is a marked gradient of electron temperature during the course of the day.

Subsequent theoretical studies clarified the parts played by the different processes determining the main features of the distribution of plasma temperatures. One very important conclusion /172/ was that thermal conduction is responsible for the fact that T_e

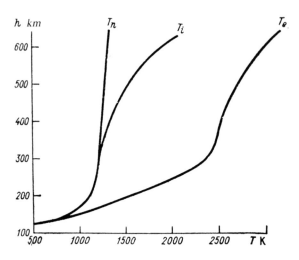

Fig. 2.2. Profiles of T_e and T_i during the daytime in
summer /146/, and the calculated T_n temperature profile using the MSIS model.

$> T_n$ at high altitudes, wherever thermal heating processes become negligible. It was
also established /164/ that a large number of photo-electrons, formed by photoionization, leave the upper atmosphere and enter the magnetosphere along the magnetic lines
of force. At high altitudes, where processes involving energy dissipation are absent,
high-energy electrons heat up the electron gas, and the heat is returned to the
ionosphere by conduction along the force lines of the magnetic field.

At low and middle latitudes the principal source of heat for both neutral and ionized
components is UV solar radiation /86/. At high altitudes the electric fields of
magnetospheric origin serve as the source of Joule heat, which may be the major
temperature-controlling factor in these regions /117, 151/.

The heating sources for the electron gas may either be local or non-local. Local
sources involve photoionization of the neutral atmosphere by short-wave solar radiation. The kinetic energy of the photo-electrons formed is much higher on the average
than the kinetic energy of the electron gas in the environment. These high-energy
photo-electrons give up their excess energy by elastic collisions with electrons, ions and
neutral particles, and also in the course of inelastic collisions, mainly with neutral
particles. The energy losses which accompany elastic collisions result in the heating of
the respective components. Since the amount of the energy transferred during a
collision depends on the mass of the "partner", the elastic collisions with the
surrounding electron gas will be more effective in this respect than collisions with
neutral particles, and it is principally the electron gas which is heated.

Several difficulties are encountered when attempting to compute the heat function of
the electron gas; for example, the effectiveness of energy transfer to the electron gas
during the deactivation of the excited neutral particle by collision is not known.
Moreover, electrons formed at a given site may well lose their energy at a different site.
As mentioned before, they are able to move along the magnetic lines of force and

produce heating at a magnetically conjugated point. Making allowances for such a source of non-local heating is in itself a highly complex task.

In calculations, the rate of heating due to the photoionization by solar radiation may be assumed to be in agreement with /280/.

Cooling of the electron gas may be produced both by elastic and inelastic collisions with neutral atmospheric components, and by Coulomb-type interaction with the ions /120, 217, 279, 21/. Thermal conduction produces substantial changes in the profile of the electron temperature as a result of the effects of heat sources and cooling of the electron gas. Heat transfer in the electron gas takes place along magnetic lines of force up to altitudes of about 80 km. The effect of thermal conductance normal to the magnetic lines of force is negligible /91/. In the upper part of the ionosphere thermal conductance depends solely on electron-ion collisions. The coefficient of thermal conductance here is the same as in a fully ionized gas /83, 85, 164/; it strongly depends on T_e and is independent of n_e.

When calculating ion temperatures at altitudes of up to 500–600 km it is permissible to neglect thermal conduction in the ion gas. The time constant of ion temperature variation is determined by charge transfer processes between ions and neutral particles and by elastic collisions with electrons /84/.

Figure 2.3 shows the results of calculations of the profiles of T_i /85/ for various constant values of the electron temperature T_e, and for the temperature of the neutral atmosphere, $T_n = 1000°K$. Below 300 km the heating of the ions by the electron gas is minor, and $T_e \approx T_n$. At higher altitudes, however, the cooling of neutral particles is ineffective and the ion temperature approaches the electron temperature. For a review of the kinetic problems of thermal and super-thermal electrons in the upper atmosphere see /43/.

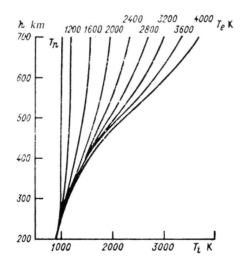

Fig. 2.3. Profiles of T_i calculated /85/ for various constant values of T_e, at $T_n = 1000°K$.

CHAPTER 2

2.2 THE FUNDAMENTAL SET OF EQUATIONS

We shall now present the equations which may be taken as the starting point for calculations in the $F2$-region, and a method for solving the system of equations obtained. As we have seen, the distribution of the electron concentration in the $F2$-region can be described by the continuity equation which, to a diffusional approximation, is written as follows:

$$\frac{\partial n_i}{\partial t} = \frac{\partial}{\partial z}\left[D_{in}\sin^2 I\left(\frac{\partial n_i}{\partial z} + An_i\right) - wn_i\right] - \beta n_i + q \qquad (2.19)$$

Here D_{in} is the diffusion coefficient of O^+ ions through neutral particles,

$$A = \frac{T_e}{n_e T_i}\frac{\partial n_e}{\partial z} + \frac{m_i g}{kT_i} + \frac{1}{T_i}\frac{\partial(T_e + T_i)}{\partial z}$$

where

$$n_e = [O^+] + [NO^+] + [O_2^+]$$

and T_e and T_i are the electron and ion temperatures, respectively, and

$$w = v_{nx}\sin I \cos I \cos D - v_{ny}\sin I \cos I \sin D$$

is the vertical velocity of the drift induced by thermospheric winds, v_{nx}, v_{ny} are the meridional and the zonal velocity components of the thermospheric wind, and I and D are the dip and inclination angles of the magnetic field of the Earth respectively, at the point in question; $\beta = \gamma_1[N_2] + \gamma_2[O_2]$ is the loss coefficient, and q is the production rate of the O^+ ions by photoionization.

Thus, all the above parameters must be known in order to calculate the electron concentration distribution. Depending on the task in hand, some of them may be available, either from experimental results or after making certain assumptions. However, as far as the overall problem of calculating the $F2$-region is concerned, it is necessary to know the values of all these parameters under well defined conditions. As a result, we have to solve the problems of choosing a suitable model of the neutral atmosphere, and calculating the photoionization and recombination rates, and plasma temperatures and velocities of the thermospheric wind. If the thermospheric model, the solar UV flux and the ionization and absorption cross-sections are all known, the photoionization rate of O, O_2 and N_2, may be represented by the equation (2.5), rewritten as follows:

$$q_j(z) = n_j(z)\sum_{k=1}^{m} I_{0k}\sigma_{ijk}\exp\left\{-Ch(\chi)\sum_{j=1}^{3}\sigma_{ajk}n_j(z)H_j(z)\right\}$$

where the integration over the spectrum has been replaced by summation over m

spectral intervals; the integral for the calculation of the number of particles in the column is represented as

$$\int_z^\infty n_j dz = n_j(z) H_j(z),$$

where $H_j(z)$ is the atmospheric scale height for the j-th component at the altitude z.

If the concentrations of O_2 and N_2 and the rate constants γ_1 and γ_2 are known, the loss coefficient β may be directly calculated: $\beta = \gamma_1[N_2] + \gamma_2[O_2]$.

The distribution of the electron temperature, T_e, may be determined by solving the energy equation

$$\frac{\partial T_e}{\partial t} = \frac{2}{3kn_e}\left[\frac{\partial}{\partial z}\left(K_e \sin^2 I \frac{\partial T_e}{\partial z}\right) + Q_e - L_e\right] \tag{2.20}$$

where K_e is the thermal electron conductivity, Q_e is the thermal electron heating rate, and L_e is a term correcting for energy of the electrons. The values of these parameters may be found in the publications cited in para. 2.1.3; for a formula giving the ionic temperatures see /84/.

The meridional and zonal components of the velocity of thermospheric wind may be determined by solving a set of two equations of motion (2.18). Assuming that $u_i = 0$ and that O^+ is the principal ionic species, these equations may be rewritten as follows:

$$\frac{\partial v_{nx}}{\partial t} = \frac{\eta}{\rho}\frac{\partial^2 v_{nx}}{\partial z^2} - \frac{v_{nx}}{\rho}\sum_j n_i\mu_{ij}v_{ij} + 2v_{ny}\Omega\sin\varphi - \frac{1}{\rho}\frac{\partial p}{\partial x}$$

$$\tag{2.21}$$

$$\frac{\partial v_{ny}}{\partial t} = \frac{\eta}{\rho}\frac{\partial^2 v_{ny}}{\partial z^2} - \frac{v_{ny}}{\rho}\sum_j n_i\mu_{ij}v_{ij} + 2v_{nx}\Omega\sin\varphi - \frac{1}{\rho}\frac{\partial p}{\partial y}$$

Clearly, in mathematical terms, the calculation of diurnal variations of electron concentration is reduced to solving the set of four parabolic second order differential equations (2.19) to (2.21). This involves the knowledge of both the initial and the boundary conditions. Since the calculation of diurnal variations of these parameters is tantamount to a search for a periodic solution, the starting conditions may be chosen fairly arbitrarily, and the computation continued until a constant solution is obtained. For the particular problem presented above this amounts to $1\frac{1}{2} - 2$ days of real time. In practice this time period may be shortened by starting the computation at about noontime, and taking the initial conditions as the ones obtained by solving the equations for input parameters corresponding to that time. This is based on the fact (cf. para. 4.2) that the $F2$-region is in a quasi-stationary state around noon.

Our aim is to calculate the distribution of electron concentration in the $F2$-region. Consequently we shall limit our considerations to the altitude of 600 km, thus enabling us to neglect the contribution of H^+-ions, at least for the middle and the high levels of solar activity, without causing a significant error. Since the problem is to be solved for a limited range of altitudes, we must stipulate boundary as well as initial conditions. Clearly, at the lower boundary the following conditions may be accepted in the diffusion equation: the O^+-ion is in photochemical equilibrium; in the thermal conductivity equation $T_e = T_n$; and, for thermospheric winds, $V_{nx} = V_{ny} = 0$.

Let us now consider the upper boundary conditions. All four equations are of the same type as before. We shall begin with some general remarks. The upper boundary condition may be given in three different forms: the function itself (for conditions of the 1st type), the derivative of the function (for conditions of the 2nd type) and a combination of these functions (for conditions of the 3rd type). Although all three conditions are equivalent, from the mathematical point of view, the conditions of the 1st type even being preferred over the other two, since they may be given accurately by using approximation by differences, from the physical point of view the conditions of the 1st type are to be treated with caution. Thus, for instance, an erroneous value for the concentration at the upper boundary and an uncoordinated selection of the system of aeronomic parameters may considerably distort the true variation of the electron concentration with altitude, and spurious fluxes, which reflect the gradients of the distorted distribution of this concentration over the latitude, may appear in the regions under study. Other difficulties of a physical nature may also arise if the upper boundary concentration is given /45/. The correct procedure, in physical terms, is to stipulate a flux as the upper boundary condition.

An exact stipulation of the flux of O^+-ions at the upper boundary involves a discussion of the mechanism of ionosphere-protonosphere interaction, presented in /44, 58, 108, 211, 220, 250/. It follows from these studies that there is an interchange of fluxes between the ionosphere and the protonosphere, and that the fluxes are one order of magnitude smaller than the so-called critical flux /165/ – for the daytime $F2$-region this is about $2 \cdot 10^9$ $cm^{-2} \cdot sec^{-1}$. Rates of motion of plasma, determined experimentally by the incoherent scatter method /147/, indicate that a typical flux is $5 \cdot 10^7$ $cm^{-2} \cdot sec^{-1}$ in the daytime and $3 \cdot 10^7$ $cm^{-2} \cdot sec^{-1}$ at night. However, these fluxes, well below the critical level, do not significantly affect the distribution of electron concentration in the $F2$-region in the daytime.

The role played by these fluxes will be discussed again in the text that follows. Suffice it to note that if the calculations of the $F2$-region are carried out to the first approximation, we may postulate a downward flux of the order of $5 \cdot 10^7$ to 10^8 $cm^{-2} \cdot sec^{-1}$, which is either constant throughout the day or varies in a stepwise fashion. A more accurate expression or the flux at the upper boundary z_1 may be obtained by integrating the diffusion equation between the altitude z_1 and infinity /274/:

$$\frac{\partial N_i}{\partial t} = \phi_i(z_1) + \int_{z_1}^{\infty} q \, dz - \int_{z_1}^{\infty} \beta n_i dz. \tag{2.22}$$

It is assumed that the flux is zero at infinity, and that the total number of particles in the column is

$$N_i = \int_{z_1}^{\infty} n_i dz.$$

In physical terms this expression indicates that a change in the number of particles in the column about the z_1-level takes place as a result of flux across the z_1-level, and is also caused by a source of ionization and recombination.

The boundary condition in the form given in (2.22) leaves the upper boundary free, allowing plasma to flow across it during the redistribution of the plasma concentration in the region in question. Since at high altitudes the processes of plasma formation and recombination are not very effective, condition (2.22) may be simplified as

$$\phi_i(z_1) = \frac{\partial}{\partial t} \int_{z_1}^{\infty} n_i dz = \frac{\partial}{\partial t} [n_i(z_1)H_p(z_1)],$$

where $n_i(z_1)$ and $H_p(z_1)$ are the electron concentration and the scale height of the plasma at the z_1-level respectively.

In calculating wind velocities, the boundary conditions may be selected on the assumption that at high values of z, caused by the high values of the viscosity coefficient, wind velocities do not appreciably change with altitude. Accordingly, the upper boundary conditions may be written as

$$\frac{\partial v_{nx}}{\partial z} = \frac{\partial v_{ny}}{\partial z} = 0$$

In order to solve the thermal conductance equation, the thermal flux at the upper boundary must be known. According to the data obtained by the method of incoherent scatter, above the $F2$-region maximum the electron temperature increases with altitude, indicating the presence of a flux directed from the magnetosphere towards the $F2$-region.

The electron temperature gradient in the altitude regions in which thermal convection is the dominant factor may be used to calculate the flux by an empirical method. Evans /145/ studied the seasonal diurnal variations of the flow caused by solar activity. The results of the observations made during the years 1964–1968 allowed the empirical values of the day-time flux during the winter (January–March, October–December) and during the summer (April–September) to be deduced as a function of the solar activity level (value of the index $F_{10.7}$ averaged over six months):

$$\phi_{summer} = 3.65 \cdot 10^7 \bar{F}_{10.7} \ eV/(sec \cdot cm^2)$$

$$\phi_{winter} = 5.4 \cdot 10^7 \bar{F}_{10.7} \ eV/(sec \cdot cm^2)$$

The daytime fluxes are twice as high as the nighttime winter ones, and 5–10 times as high the summer ones. According to /281/, variations of the flux with latitude are insignificant.

Now that the equations have been presented in this form and the boundary conditions determined, the task has been defined, and the next stage is to solve it. Analytical methods are not always convenient to use in the calculations. Since solutions of this type are only available for conditions in simplified models, even the solution of the diffusion equation alone is a fairly complex task /62, 126, 166/. Consequently, only a numerical solution of the set of four interrelated equations (2.19) to (2.21) is possible. However, since there is no theory which gives numerical solutions of nonlinear parabolic equations of this kind, the correctness of any solution, if at all possible, will always be in doubt. Criteria such as stability and convergence of any given method of solution do not invariably indicate that the solution is correct /53/. If the results of the calculations are compared with experimental results, and they do not correspond, this may either be due to an incorrect method of calculation or to erroneous selection of parameter values. One possible approach to this problem is to compare the numerical solutions with analytical results obtained with simple models, and to use the results of this comparison to decide which numerical method to choose.

Problems of this type are frequently solved by the method of finite differences /67/, in which the starting equations are written down in the form of finite differences, which are then solved. Other approaches are also known /218/. The method of finite differences involves special difficulties when applied to the diffusion equation, and for this reason will be considered separately.

Let us rewrite equation (2.19) in the form:

$$\frac{\partial n}{\partial t} = \frac{\partial}{\partial z}\left[D\left(\frac{\partial n}{\partial z} + An - \frac{wn}{D}\right)\right] - \beta n + q. \qquad (2.23)$$

This equation is of a quasi-linear type, since all its coefficients are functions of altitude and time, the former dependence being particularly pronounced; in particular, the increase in the coefficient D with altitude is exponential and this is the principal source of the difficulties encountered in attempting a numerical solution.

The different procedures for the approximation of equation (2.23) must satisfy a number of definite conditions:

1) The procedure must be stable both with respect to the round-off errors and the initial data;

2) The steps involved in the integration with respect to coordinates and to time should be chosen according to physical considerations and the degree of the accuracy required, and not be imposed by the specific features of the procedure chosen;

3) The procedure must be suitable for solving a large number of problems in the particular class and should furnish an adequate description of the physical effect under study; and

4) The procedure must ensure efficient utilization of computer time.

Practical experience has shown that the accuracy and stability of the method of finite differences depend on the specific form of the approximation equation. In general, equation (2.23) is re-written as follows:

$$\frac{\partial n}{\partial t} = a_1 \frac{\partial^2 n}{\partial z^2} + a_2 \frac{\partial n}{\partial z} + a_3 n + a_4 \qquad (2.24)$$

for which the differential problem is then posed /67/. The use of explicit difference methods is undesirable, since the stability condition $\Delta t < (\Delta h)_2/2D$, where Δt and Δh are the time and the altitude steps, respectively, imposes serious constraints on the time step. Since the diffusion coefficient D at the altitude of 600 km is approximately 10^{12} cm^2/sec, the time step corresponding to an altitude step of 5–10 km is less than 1 second, which is intolerable for practical purposes.

If a fully implicit method of differences is used, and equation (2.24) is approximated indirectly, the estimated time step will be:

$$\Delta t < 1\bigg/\left(\frac{\partial(DA)}{\partial z} - \beta\right)$$

where Δt may be of the order of several tens of seconds, which may be of practical interest.

The theory of difference methods has been most fully developed for parabolic equations in the form /67/:

$$c\frac{\partial u}{\partial t} = \frac{\partial}{\partial x}\left(k\frac{\partial u}{\partial x}\right) + pu + f. \tag{2.25}$$

Depending on the manner in which equation (2.23) is reduced to the form (2.25), the sign of the coefficient p may change – negative at low altitudes owing to the highly effective dissipation processes, and positive at higher altitudes. An additional difficulty is that in the region where p is positive, the solution becomes unstable.

To obtain stability throughout the region in question, a suitable selection of the integration steps is necessary, in which the time step is usually reduced, while the altitude step remains unchanged. Mikhailov and Ostrovskii /53/ gave a comparative analysis of the various procedures employed in solving the diffusion equation. The comparison of the analytical with the numerical solutions made it possible to evaluate the accuracy of calculations made by the various methods. It was shown that when fast computers are used, a fully implicit procedure may ensure a high accuracy in calculation, making it suitable for use. However, it involves checking the stability conditions and selecting Δt values at each time step. Another acceptable procedure is the well known method of Gershengorn /10/, using large time steps while preserving the requisite accuracy, which enables a stable calculation to be carried out. The relevant equation is obtained from (2.23) by an exchange of variables, $n = y\mu$, where

$$\mu = \exp\left(-\int_0^z\left(A - \frac{w}{D}\right)dz\right) \tag{2.26}$$

As a result, the equation is reduced to the form/(2.25)

$$\frac{\partial(y\mu)}{\partial t} = \frac{\partial}{\partial z}\left(D\mu\frac{\partial y}{\partial z}\right) - \beta\mu y + q \tag{2.27}$$

when the coefficient $p = -\beta\mu$ is negative for all heights.

Let us select a uniform grid with the coordinate step Δh and the time step Δt and introduce the grid function μ_i^j. The continuous equation (2.27) may then be written in the form of differences as follows /67/:

$$\frac{u_i^{j+1}\mu_i^{j+1} - u_i^j\mu_i^j}{\Delta t} = \frac{1}{\Delta h}\left[\bar{\lambda}_{i+1}\frac{u_{i+1}^{j+1} - u_i^{j+1}}{\Delta h} - \bar{\lambda}_i\frac{u_i^{j+1} - u_{i-1}^{j+1}}{\Delta h}\right] -$$

$$-R_iu_i^{j+1} + q_i \tag{2.28}$$

$$\bar{\lambda}_{i+1} = \frac{\lambda_{i+1} - \lambda_i}{2},\ \bar{\lambda}_i = \frac{\lambda_i + \lambda_{i-1}}{2},\ \lambda = D_\mu,\ R = \beta\mu,$$

$$i = 1, 2, ..., N-1, j = 0, 1, ...$$

Here the parameters λ, R and q, are usually taken on the $(j+1)$-th time layer. In cases where the time step is large or the parameters vary rapidly, the average values between consecutive nodes may be taken.

The difference equation (2.28) for the nodes of the grid results in a system of four algebraic equations giving the values of function u_i in the $(j+1)$-th time layer, assuming that the values of u_i in the j-th layer are known:

$$A_i u_{i-1}^{j+1} - C_i u_i^{j+1} + B_i u_{i+1}^{j+1} = -F_i; \quad i = 1, 2, \ldots, N-1,$$

$$A_i = \bar{\lambda}_i/\Delta h^2; \quad B_i = \bar{\lambda}_{i+1}/\Delta h^2,$$

$$C_i = \bar{\lambda}i/\Delta h^2 + \bar{\lambda}_{i+1}/\Delta h^2 + R_i + \mu_i^{j+1}/\Delta t,$$

$$F_i = \frac{\mu_i^j}{\Delta t} u_i^j + q_i.$$

The system may be solved when the two boundary conditions, at $i = 0$ and $i = N$ are known. Systems of linear equations of this type are now usually solved by the highly effective "run" method /67/. The first step is to calculate the "run" coefficients, M and Z:

$$M_{i+1} = \frac{B_i}{C_i - A_i M_i}, \quad Z_{i+1} = \frac{A_i Z_i + F_i}{C_i - A_i M_i}, \quad i = 1, 2, \ldots, N-1.$$

In this particular case the initial values are: $M_1 = 0$; $Z_1 = n_0$, where n_0 is the lower boundary concentration. The values themselves may then be found.

$$u_i = M_{i+1} u_{i+1} + Z_{i+1}, \quad i = 0, 1, \ldots, N-1. \tag{2.29}$$

The value of u_N is determined by the upper boundary condition which, for the situation considered here, may be written as

$$u_N = [Z_N(\lambda_N + \lambda_{N-1}) - 2\Delta h\phi]/[(\lambda_N + \lambda_{N-1})(1 - M_N)]$$

Such a relationship is obtained by simultaneous consideration of equation (2.29) and the expression for the flux, which has the value of ϕ at the upper boundary. Equations for thermospheric winds and thermal conductivity may be presented in a similar manner.

Despite the fact that a stable count is obtained by this method, the solution of the diffusion equation meets with purely computational difficulties. Thus, at altitudes below 200 km, owing to the decrease in the coefficient of diffusion, the values of μ may become too large to be handled by the computer, even at low drift velocities. In certain cases it may be sufficient to normalize the magnitude μ. However, the diurnal rate of the vertical drift is very variable, ranging from -20 m/sec in daytime to $+60$ m/sec or more at night; this produces large changes in the value of μ so that in practice the selection of the normalization procedure is far from being a simple matter.

With respect to the zone adjacent to the layer maximum, the following method may be adopted. At a distance below the maximum, which is equal to one atmospheric scale height, for example, a zero drift may be inserted in the expression for μ. From the mathematical point of view, such a procedure settles the problem of large μ-values, while being justified from the physical point of view as well, since the photochemical processes predominate at low altitudes. Normalization, as well as causing a zero vertical drift at a given altitude, also makes it possible to calculate the diurnal variations of the F2-region under various helio-geophysical conditions.

The system under consideration consists of several interrelated equations, and the most common method for their solution involves successive "runs" with iterations. Thus, the equations are successively solved, including a number of iterations (usually two or three) for each time step, which ensures the consistency of the calculated parameters. Iterations are mandatory, since when the equations are solved in succession, some of the parameters to obtain the coefficients are taken from the preceding time layer, which may give rise to errors, especially if the time steps are large.

The choice of the step size is determined both by the requirements for stability and for the accuracy in the calculation, and also by physical considerations. If the non-reduced diffusion equation in the form (2.23) is solved by implicit methods, stability can only be ensured /298/ if the time step is shorter than the characteristic time of the fast process. At high altitudes, when stability is impaired, the diffusion process is the fast one, and the characteristic time at an altitude of 600 km is $H^2/D \approx 10$ sec, in agreement with the estimated Δt for the implicit method described above. If, on the other hand, the equation is converted to its special form (2.27), the resulting method is stable whatever the step size, and the selection of the time step is determined by physical considerations. Thus, if a large step size, which exceeds the characteristic times of the main processes, is chosen, information is lost, as was shown by Waldman /298/, so that it is preferable in such a case to solve a set of stationary equations.

In the case of the layer maximum in which the characteristic times of the diffusion process, H^2/D, and the recombination process, $1/\beta$, are approximately equal (about 1½ hours), the time step chosen should be shorter than this. It is found in practice that the count gives satisfactory results from $\Delta t = 10$ to 30 minutes. However, the time step must be reduced to a few minutes around sunrise and sunset, in view of the fast variations which then occur in the neutral atmosphere and in the photoionization rate.

2.3 THE EFFECTS OF VARIOUS PHYSICAL PROCESSES ON F2-REGION IONOSPHERE FORMATION

The F2-region is formed as a result of three simultaneous processes: production, transport and recombination of ions and electrons. Each process in turn involves a number of individual physical processes, each affecting the electron concentration distribution.

Thus, for instance, plasma transport is related to diffusion in a gravity field, with thermospheric winds and electric fields, while recombination takes place at the final stage of ion conversion by a series of ion-molecule reactions due to dissociative recombination of molecular ions. The role played by the individual processes in F2-region formation will be discussed in this section. We shall also present relationships by means of which the parameters of the layer maximum can be calculated.

2.3.1 Relationships Involving the $F2$-Layer Maximum. A more complete description of the behavior of the $F2$-region is only possible if between two and four parameters are specified: electron concentration at the layer maximum, the altitude of the layer, and data on the shape of the layer above and below the maximum. Solutions of the diffusion equation indicate that certain relationships are valid in the layer maximum; these relate the electron concentration and the layer height to the scale height H, and to parameters such as the ion formation rate q, the loss coefficient β, and the diffusion coefficient D, which are employed in the theory of ionosphere formation. Calculations performd by Rishbeth and Barron /65, 246/ yielded relationships of the following form:

$$\beta_m \approx 0.6 D_m/H^2, \; n_c^m \approx 0.75 q_m/\beta_m, \tag{2.30}$$

where H is the scale height of neutral atomic oxygen. These relationships are approximate, and the coefficients are inaccurate. A more accurate estimate of these coefficients may be made with the aid of an analytical solution.

The physical meaning of expressions in (2.30) is obvious. The second expression means that at the $F2$-layer maximum the rate of ion formation processes q is somewhat faster than the rate of recombination processes, βn_c, while in the $F1$- and E-regions they are equal. This difference is caused by the more pronounced effect of diffusion, which is still a minor factor in the lower part of the $F2$-region, but which becomes predominant above the $F2$-layer maximum. As a result, ions and electrons formed above the layer maximum mostly fail to recombine at their site of formation, but descend and recombine at altitudes below the layer maximum: just below the maximum $q \lesssim \beta n_c$, while at maximum and in the upper part of the $F2$-region, $q > \beta n_c$.

The first expression in (2.30) shows that in the layer maximum the characteristic times of recombination $1/\beta$ and of diffusion at a distance of one scale height H are approximately H^2/D, and both are roughly equal. This may be understood from the following considerations. For the sake of simplicity, we shall consider the case of zero vertical drift ($w = 0$): the coefficient $D \sim e^{z/H}$, increases, while the coefficient $\beta \; e^{-z/H\beta}$ decreases with increasing altitude. Accordingly, in the specific altitude range, which corresponds to the minimum of the reciprocal of the resulting characteristic lifetime of the particles τ (according to the theory of scale analysis, τ in (2.19) is given approximately by $1/\tau \approx \beta + D/H^2$), we may expect to be able to locate the maximum concentration, n_c, i.e., the site of the layer maximum. Clearly, $1/\tau$ will be at minimum when $\beta/H_\beta = D/H^3$, equivalent to (2.30), since $H_\beta \approx H/1.75$ holds true.

We shall now turn to the analytical solution of the diffusion equation and specify the conditions of formation of the layer maximum in more detail. For the sake of simplicity, we shall consider an idealized model of the daytime stationary $F2$-region, which is formed in an isothermal neutral atmosphere at $w = 0$. Introducing $z = h - h_0$, where h_0 is the selected altitude, we have

$$q = q_0 \exp(-a_1 z), \; \beta = \beta_0 \exp(-a_2 z), \; D = D_0 \exp(a_3 z),$$

where a_i is the reciprocal of the scale height for the respective atmospheric component, e.g. $a_1 = 1/H$, $a_2 = 1/H_\beta$ and $a_3 = 1/H_D$. This is based on the conclusion reached in para.

2.1.1 to the effect that at altitudes of $F2$-region the rate of the ion formation q is proportional to the concentration of atmospheric particles. The diffusion equation may then be written in the form:

$$\frac{d^2 n_e}{dz^2} + (a_3 + a_4)\frac{dn_e}{dz} + a_3 a_4 n_e + \frac{q_0}{D_0}x^\gamma - \frac{\beta_0}{D_0}n_e x = 0, \qquad (2.31)$$

where

$$x = \exp[-z(a_2 + a_3)], \quad \gamma = (a_1 + a_3)/(a_2 + a_3)$$

and $a_4 = \frac{1}{2}H$ is a magnitude which is inversely proportional to the plasma scale height.

An analytical solution of equation (2.31) is available at constant a_i and zero flux at the upper boundary /165/. According to /29/, this expression becomes particularly simple if $z = 0$ is selected for a level at which the relationship

$$\beta_0 = D_0(a_2 + a_3)^2. \qquad (2.32)$$

is valid. We then have

$$n_e = \frac{q_0}{\beta_0}\left\{ \Gamma(\lambda)\Gamma(\lambda - v)x_\lambda\left[x_{-\lambda}\sum_{k=0}^{\infty} \frac{x^k}{\Gamma(1+k)\Gamma(1+k-v)} - \right.\right.$$

$$\left.\left. - \sum_{k=0}^{\infty} \frac{x^k}{\Gamma(1+\lambda+k)\Gamma(1+\lambda-v+k)}\right]\right\} \qquad (2.33)$$

where

$$\lambda = (a_1 + a_3 - a_4)/(a_2 + a_3), \quad v = (a_3 - a_4)/(a_2 + a_3)$$

According to this solution, the layer maximum is given, to a close approximation /30/, by

$$x_m^{-\lambda} = \frac{\gamma\Gamma(2-v)[1 + (\gamma+1)x_m/\gamma(\lambda+1)(\lambda+1-v)]}{(1-v)(\gamma-\lambda)\Gamma(1+\lambda)\Gamma(1+\lambda-v)[1+(\gamma-\lambda+1)] \times} \approx$$

$$\times x_m/(\gamma-\lambda)(1-v)]$$

$$\approx \frac{\gamma}{(\gamma-\lambda)(1-v)} = \frac{a_1 + a_3}{a_4(1-v)} \approx 4 \qquad (2.34)$$

The right-hand side of this expression is approximate, and is based on the assumption that the ratio between the expressions in square brackets and the values of $\Gamma(1 + \alpha)$, where $0 < \alpha < 1$, is approximately equal to unity.

In particular, under standard conditions, when $1/H = a_1 = a_3 = 2a_4 = a_2/2$; having found that $\lambda = 0.5$, $\gamma = \frac{2}{3}$ and $v = \frac{1}{6}$, we obtain $x_m \approx 0.066$. More accurately, if $a_2 H = \frac{28}{16}$, we have $\lambda = 0.546$, $\gamma = 0.728$, $v = 0.164$, and we find, from (2.34) and (2.33)

$$x_m = 0.06 \text{ and } n_e = 1.86 q_0/\beta_0 \qquad (2.35)$$

Let us compare the expressions thus obtained with the approximate relationships (2.30). Since

$$q_m = q_0 x_m^{1/3}, \; \beta_m = \beta_0 x_m^{2/3}, \; D_m = D_0 x_m^{-1/3} T,$$

by taking (2.32), and the first relationship in (2.35) into consideration, we obtain:

$$\frac{\beta_m H^2}{D_m} = \frac{\beta_0}{D_0(a_2 + a_3)^2} 9 x_m = 0.54 \tag{2.36}$$

This is close to the first relationship in (2.30). On the other hand, the second relationship in (2.35) may be rewritten as follows:

$$n_e^m = 1.86 \frac{q_m}{\beta_m} x_m^{1/3} = 0.73 \frac{q_m}{\beta_m} \tag{2.37}$$

which is almost identical to the second relationship in (2.30). These relationships may be conveniently employed in solving a number of problems.

In order to be able to calculate the parameters of the $F2$-layer maximum which are a function of q_m, β_m and D_m, from an atmospheric model, the reference parameters of the upper atmosphere cannot be those at the $F2$-layer maximum, but must be at some other level, z_1. This question will now be discussed.

Since at any level of isothermal atmosphere the expressions Dx/β and $qx^{1/3}/\beta$ are invariant, we find from (2.36) and (2.37) for $z = z_1$:

$$\beta_1 H^2 / D_1 x_1 = 0.54/x_m \text{ and } n_e^m = 0.73 q_1 / \beta_1 (x_1/x_m)^{1/3}$$

We now pass from the magnitudes employed in the theory of formation of the ionosphere to the concentrations of particles in the atmosphere. This is done by expressing the magnitudes $q = [O]j_o$ and $D = d[O]^{-1}$, in terms of $[O]$, where $j_o = 3 \cdot 10^{-7} I_{150} I_o \sec^{-1}$ and $d = 1.38 \cdot 10^{19} \sqrt{T/1000} \text{ cm}^{-1} \cdot \sec^{-1}$. We then obtain the following equations *in lieu* of (2.36) and (2.37)

$$x_1/x_m = \frac{H^2 \beta_1 [O]_1}{0.54 d}, \tag{2.38}$$

$$n_e^m = \frac{0.73 j^o}{(0.54 d)^{1/3}} \left[\frac{[O]_1^2 H}{\beta_1} \right]^{2/3} \tag{2.39}$$

It follows from the definition of x that the altitude, $h_m - h_1 = z_m - z_1 = \frac{H}{3} \ln(x_1/x_m)$ may be calculated from (2.38). At a given level of solar activity, j_o and d are constant. It is important to note that the altitude, h_m, and the electron concentration n_e^m at a given level, h_1 (e.g., $h_1 = 300$ km) depend solely on the values of H, $[O]$ and β at that level, which is readily allowed for in the atmospheric model.

We shall now analyze the solutions (2.38) and (2.39) just obtained. The explicit expression for the layer maximum height is

$$h_m - h_1 = \frac{H}{3}[\ln(\beta_1(O)_1) + \ln(H^2/0.54d)] \tag{2.40}$$

Both right-hand terms of this equation are functions of the temperature T, by way of H; moreover, changes in T or other heliogeophysical parameters result in changes of β_1 and $[O]_1$.

It is seen from (2.40) that the altitude, h_m, is a function of the product of β_1 and $[O]_1$, both factors being equally important. According to (2.39), the electron concentration n_e^m is a function of the ratio $([O]_1^2 H/\beta_1)^{2/3}$. Thus, a variation of $[O]_1$ causes an in-phase change in the h_m and n_e of the layer maximum, while a change in β_1 causes these parameters to change in counterphase.

Since the ratio $[O]_1^2 H \beta_1$, remains approximately constant when the atmospheric temperature alone changes, the variation in n_e^m is insignificant, while the altitude variation h_m, which is directly proportional to H, is considerable. In contrast, a change in the intensity of short-wave radiation only causes n_e^m to change, while the altitude h_m remains unchanged. It is interesting to note that if $F \approx 150$, and $T_{300} \approx 1150°K$ so that $H \approx 65$ km, we have

$$\Delta h_m = 50\Delta \lg[O] + 50\Delta \lg\beta. \tag{2.41}$$

We may note that the nature of the variation of n_e^m and h_m with j_o and the density of the atmosphere is in agreement with similar conclusions of the simple layer theory when z_o is given, but these are inapplicable to the F2-region from the point of view of the dependence on z_o.

It is even easier to obtain a stationary solution for nighttime conditions for an idealized model of the atmosphere with $w = 0$ /33, 165/:

$$n_e^m = -\frac{2H\phi_\infty}{D_1} v' \left(\frac{h_m - h_1}{H}\right) \text{ and } \frac{h_m - h_1}{H} \approx 0.55 \tag{2.42}$$

where ϕ_∞ is the ion flux from the protonosphere to the ionosphere and h_1 is the height at which $\beta_1 = D_1/4H^2$, v' is a universal altitude function, similar to Chapman's function in the simple layer theory. In this situation, if the atmospheric temperature is given, n_e^m is solely determined by the magnitude ϕ_∞, while the height h_m is larger than h_1 by about one-half of the altitude scale height. The altitude of the stationary F2-region, at night and by day, is a function of $H/3 \ln (\beta_1[O]_1)$.

Equations (2.39) and (2.40) may also be used to solve the converse problem: to estimate the atomic oxygen content and the loss coefficient if the altitude and the electron concentration in the maximum F2-layer are known. The direct and the converse relationships for real conditions will be given below.

The above analysis is only valid for a relatively simple idealized case. If the work is based on a real model of the thermosphere, numerically more accurate expressions for the coefficients of diffusion and recombination and effects connected with vertical drifts should be obtained with the aid of a computer. Even so, these solutions broadly resemble (2.39) and (2.40). Thus, using the thermospheric MSIS model /177/ for $F_{10.7} = 150$, diffusion coefficient as given by /91/ and values of γ_1 and γ_2 by the expression $\beta = \gamma_1[O_2] + \gamma_2[N_2]$ according to /155, 215/, we obtain the following relationships between the parameters of mid-latitude F2-region around midday and the atmospheric parameters at $h = 300$ km with the vertical drift w:

$$\lg n_c^m = 1.08\lg[O] - 0.65\lg\beta + 9\cdot10^{-3}w + \lg I/I_{144} - 5.8, \qquad (2.43)$$

$$h_m = 50\lg[O] + 50\lg\beta + 1.55w + 30, \qquad (2.44)$$

where w is the drift velocity (m/sec), the upward drift being taken as positive, and I/I_{144} is the ratio between the total flux of short-wave solar radiation and its value at the solar activity level $F_{10.7} = 144$.

The presence of additional terms in (2.43) and (2.44), which do not appear in (2.39) and (2.40) is due to the fact that in the real F2-region there are vertical drifts of the plasma, induced by thermospheric winds and electric fields, which displace the layer upwards or downwards; this is reflected in the electron concentration as well. The effects of vertical drifts will be discussed in more detail in the next section. The term in (2.43) which describes the variation of the total flux of the solar radiation I is also present in (2.39) as the effectiveness of the ionization, j_o, which is proportional to I.

Equations (2.43) and (2.44) have been derived for medium and medium-to-high levels of solar activity $F_{10.7} = 150\pm30$; the results of numerical solution are within ±10 km for the altitude of the layer maximum and within ±0.1 for log n_c^m. Similar relationships are obtained for the high and low levels of solar radiation, but the values of the coefficients of $\lg[O]$ and $\lg\beta$ are slightly different.

Using the equations (2.43) and (2.44) it is possible to calculate the values of the altitude and electron concentration directly, without recourse to a computer. Table 2.2 shows the results of the calculation of $\lg n_c^m$ and h_m for a number of days in 1970 during

TABLE 2.2

Values of $\lg n_e^m$ and h_m found by using formulas (2.43) and (2.44) and by solving equation (2.19) numerically.

	lg[O]	lg β	lg I/I_{144}	ω m/sec	h_m km (2.44)	hm km (2.19)	lg n_c^m (2.43)	lg n_c^m (2.19)
6	8.942	−3.823	−0.071	−11	270	266	6.16	6.084
20	9.019	−3.612	0.064	−5	294	291	6.30	6.261
48	9.09	−3.431	0.1	1	315	313	6.23	6.351
53	9.089	−3.512	0.083	7.5	322	321	6.36	6.467
76	9.043	−3.439	−0.022	4.5	318	311	6.21	6.213
83	9.092	−3.395	0.057	−0.5	315	306	6.21	6.282
104	9.077	−3.239	0.061	−1	321	319	6.09	6.169
115	9.002	−3.289	0.009	−4.5	310	306	6.02	6.021
161	8.866	−3.311	−0.004	−1	307	314	5.90	5.913
175	8.856	−3.311	−0.013	−1	307	317	5.88	5.904
188	8.877	−3.315	0.041	−6	300	306	5.91	5.948

different seasons (the table gives the serial number for the days of the year). The calculation was made using the formulas (2.43) and (2.44), and by solving the diffusion equation (2.19) numerically. The table also contains the calculated values of lg [O], lg β and w at the altitude of 300 km, and lg I/I_{144}. It is seen that the degree of accuracy specified above is in fact achieved by using the formulas (2.43) and (2.44).

The linear relationships (2.43) and (2.44) may be written in the different form as follows:

$$\Delta lg_{c}^{m} = 1.08\Delta lg[O] - 0.65\Delta lg\beta + 9\cdot10^{-3}\Delta w + lgI_1/I_2, \tag{2.45}$$

$$\Delta h_m = 50\Delta lg[O] + 50\Delta lg\beta + 1.55\Delta w. \tag{2.46}$$

They are very useful for making quantitative estimates of the contributions of individual aeronomic parameters to the variation of the altitude and electron concentration in the $F2$-layer maximum.

Equations (2.43) and (2.44) may also be used to find the converse relationships, i.e., to find two out of three aeronomic parameters ([O], β and w) from the known values of the altitude and electron concentration (the variation with I is obvious).

Even though the mechanism of formation of the $F2$-region is different in the nighttime than in daytime (cf. para. 4.2), it is possible to find an expression resembling (2.44) for the altitude of the maximum of the former region. The aeronomic parameters, [O], β and w, at 300 km altitude are related to the altitude of the layer maximum as follows /61/:

$$h_m = 40lg[O] + 40lg\beta + 1.0w + 112 \tag{2.47}$$

According to the above, the coefficients of [O] and β are equal, but their values are lower than the coefficients in equation (2.44), owing to the fact that the temperature is lower at night. The relationship (2.47) from the results of numerical calculations is accurate to within ±10 km, and is valid during the period of time extending, roughly, from between 1 hour after sunset and one hour before sunrise.

Table 2.3 shows the values of h_m, calculated from (2.47) and those obtained by numerical solution of the diffusion equation at nighttime performed at the Millstone-Hill station. The values of lg [O], lg β and w are also given.

TABLE 2.3

Values of h_m (km) found by using (2.47) and by numerical solution of equation (2.19).

	Time, hours								
	21	22	23	0	1	2	3	4	5
lg [O]	8.85	8.78	8.70	8.6	8.59	8.57	8.56	8.59	8.78
lg β	−3.82	−3.93	−3.95	−3.88	−3.81	−3.74	−3.72	−3.71	−3.70
ω m/sec	20	26	31	42	35	27	24	20	−7
h_m (2.47)	342	338	339	350	344	339	335	332	314
h_m (2.19)	344	341	343	350	339	330	328	326	310

2.3.2 Effects of Protonospheric Fluxes and Drifts on the $F2$-Region. As mentioned in para. 2.1, dynamic processes are dominant in the $F2$-region. The plasma transport in the $F2$-region is related to the process of diffusion in a gravity field, and to thermospheric winds and electric fields. For the present purpose, we may represent the vertical component of the plasma velocity as $v_z = v_d + w$, where v_d is the rate of diffusion, and w is the rate of drift due to the neutral winds and to electric fields, these effects being treated as indistinguishable for our purposes. The expression for velocity in the ambipolar approximation may be written as follows:

$$v_z = -D_a\left[\frac{1}{n_e}\frac{dn_e}{dh} + \frac{1}{H_p}\right] + w, \tag{2.48}$$

where

$$H_p^{-1} = \frac{m_ig}{k(T_e + T_i)} + \frac{1}{T_e + T_i}\frac{d(T_e + T_i)}{dh}$$

It is apparent that, depending on the relationship between the effective scale height, $H_{eff} = \left(-\frac{1}{n_e}\frac{dn_e}{dh}\right)^{-1}$, with which the plasma is in fact distributed, and the H_p scale, describing the diffusional-gravitational distribution which the plasma tends to assume, and depending on the sign of w, the overall velocity v_z may be directed downwards or upwards. Observations made by the incoherent scatter method indicate that at heights above 500 km, the plasma flows upwards during the daytime, while below that altitude it flows downwards /146/. At the altitudes of the $F2$-region the plasma flows downwards at night.

Evans /140/ analyzed the vertical velocity of plasma observed at Millstone Hill, and proposed a mechanism for this motion.

Thermospheric winds produced by diurnal variations of regions at the maximum and minimum values of T_n and the electric fields cause the plasma to drift. For an expression giving the vertical component w of the drift rate, see para. 2.1.2

At an altitude of approximately 300 km the viscosity of the neutral gas is very high, and therefore there is little variation of the velocity of thermospheric winds with altitude; according to the measurements carried out at Jicamarca by the method of incoherent back scatter /301/, the vertical component of the electromagnetic drift also remains unchanged with altitude. It follows that in the altitude range of interest, where the role played by dynamic processes is significant, the velocity w, probably does not vary to any great extent with altitude. In this case, as follows from the expression (2.48), the effects caused by the plasma drift may be expected to occur at the layer maximum and just above it, since as the altitude increases, the diffusion velocity rapidly exceeds the drift velocity. The reason for this effect is the exponential increase of the diffusion coefficient D_a with altitude, so that the plasma distribution is solely determined by diffusion processes at high altitudes.

The diurnal variations in the temperature and in the concentrations of charged particles are responsible for variations of the pressure of the plasma in the $F2$-region, and the magnetic tube of force with its base section at a given site on the ground becomes filled by day and emptied by night /44, 58, 211, 220/. These plasma currents connect the ionosphere with the overlying protonosphere and, when studying the F2-region, may be considered as the upper boundary conditions.

We shall use simple model solutions to introduce the ionospheric-protonospheric fluxes and vertical drifts and point out the effects they produce on the altitude distribution of electron concentration.

The stationary equation of diffusion in the ambipolar approximation for the F2-region may be written as:

$$\frac{d}{dh}\left\{D_a\left[\frac{dn_e}{dh} + \left(\frac{1}{H_p} - \frac{w}{D_a}\right)n_e\right]\right\} + q - L = 0 \qquad (2.49)$$

Analytical solutions of equation (2.49) for idealized isothermal atmosphere and a constant value of H_p have been obtained /29, 48, 104, 165, 166, 304/.

However, the effect of fluxes on the $n_e(h)$ profile may also be grasped by solving a simpler equation, in which the terms related to the formation and recombination of charged particles and the drift w are neglected.

Moreover, in view of the fact that $D_a = D_0 \exp[(h - h_0)/H]$, where H is the scale height of the neutral atmosphere, which remains constant with altitude, equation (2.49) may be written as follows:

$$\frac{d^2n_e}{dh^2} + \left(\frac{1}{H_p} + \frac{1}{H}\right)\frac{dn_e}{dh} + \frac{n_e}{H_pH} = 0 \qquad (2.50)$$

The general solution of this equation is

$$n_e = A \exp(-h/H) + B \exp(-h/H_p)$$

where A and B are integration constants, which may be determined from the boundary conditions.

Two linearly independent solutions correspond to the two limits of the electron concentration distribution with the altitude. The solution of $B \exp(-h/H_p)$ is the diffusional-gravitational distribution at the plasma scale height H_p. The zero velocity v_z corresponds to this solution. At the other limit, $A \exp(-h/H)$ corresponds to the distribution of n_e with the scale height H, i.e., to the distribution of the neutral atmospheric component – in this case atomic oxygen. This is the so-called critical distribution, and the corresponding velocity v_z, and the flux

$$\phi_{cr} = n_e v_z = \frac{n_0 D_0}{H}\left(1 - \frac{H}{H_p}\right)$$

are also termed critical.

The significance of the critical flux ϕ_{cr} rests in the fact that it is the maximum possible diffusional upward flux of the plasma for the given neutral atmosphere. The relevant rate of diffusion varies exponentially:

$$v_z = \frac{D_0\exp(h/H)}{H}\left[1 - \frac{H}{H_p}\right]$$

and increases with the scale height H, unlike the concentration of the neutral component, which decreases with the scale height.

Since in the case here considered sources of plasma formation and recombination are absent ($q = L = O$), the flux remains unchanged with the altitude. Assuming that at the altitude, $h = h_0$, we have $n_e = n_0$, while the upward flux at this point is ϕ, the solution of equation (2.50) may then be written:

$$n_e = n_0 \frac{\phi}{\phi_{cr}} e^{-(h-h_0)/H} + n_0 \left[1 - \frac{\phi}{\phi_{cr}} \right] e^{-(h-h_0)/H_p}. \qquad (2.51)$$

If $\phi = 0$, we have the diffusional-gravitational distribution of the plasma over the altitude; if $\phi = \phi_{cr}$, the distribution is the critical one. If $0 < \phi < \phi_{cr}$, we obtain intermediate distributions over various scale heights depending on the flux ϕ.

Since $H_p \approx 2H$, at $\phi < \phi_{cr}$ the second term in (2.51) becomes dominant at higher altitudes, and the distribution tends to be diffusional-gravitational. As the flux ϕ approaches ϕ_{cr}, the critical distribution over the scale height H will extend into higher altitudes. If the flux is negative (downward), the distribution assumes the form of a layer, and the scale height, H_p, is preserved at high altitudes. This result shows that a distribution with a characteristic maximum is formed as a result of diffusional downward plasma transfer in the presence of plasma outflow at the lower boundary.

The discussion given above fails to take into account the formation of charged particles and their recombination, but the solution of the complete equation for the real $F2$-region displays the same typical features. The calculation of the $n_e(h)$profile for the idealized isothermal atmospheric model for various fluxes, ϕ, under daytime conditions may be found in /33, 165, 231/.

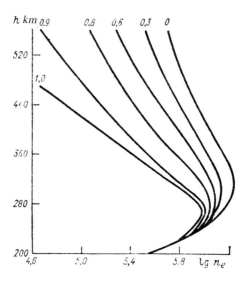

Fig. 2.4 The distribution of electron concentration in the daytime $F2$-region for differe ´ flux values at the upper boundary (ϕ/ϕ_{cr}).

Figure 2.4 shows the calculated distribution of electron concentration with altitude in a real noontime mid-latitude $F2$-region for different fluxes at the upper boundary ($\phi_{cr} = 2.8 \cdot 10^9$ cm^{-2}·sec^{-1}). At low altitudes the same solution is obtained, since ionization-recombination processes predominate, in and above layer maximum, on the other hand, the effects of fluxes become noticeable; these affect the value of h_m, concentration in the layer maximum and the scale height of the plasma distribution above the maximum. These effects of ionospheric-protonospheric fluxes have been discussed in detail by several workers /89, 163, 165, 266, 305/.

It is seen from Fig. 2.4 that the most conspicuous changes in the electron concentration profile begin at flux values close to ϕ_{cr}. Under stationary daytime conditions the critical flux for O^+-ions diffusing in a medium of atomic oxygen is about $2 \cdot 10^9$ cm^{-2}·sec^{-1}. This is much higher than the values actually observed ($\phi \approx 10^8$ cm^{-2}·sec^{-1}) /146, 147, 230/. This means that in the daytime $F2$-region the flux effects are insignificant and the plasma above the layer maximum is close to the diffusional-gravitational distribution.

As distinct from the upward flux, the descending flux is not limited, and the electron concentration in the layer maximum is proportional to the flux at the upper boundary (cf. para. 2.3.1). The concentration distribution n_e over the altitude above the layer maximum and the scale height H_p are consistent with the solution of (2.51). The altitude of the maximum, under nighttime stationary conditions, and also when the flux changes only slightly at the upper boundary, is independent of the flux ϕ and determined by the ratio between the rates of diffusion and recombination /61, 241/ (cf. para. 2.3.1).

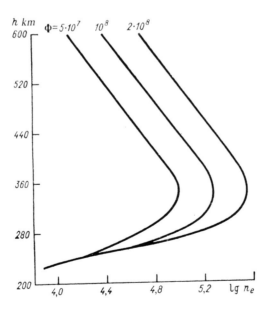

Fig. 2.5. Distribution of electon concentration under stationary nighttime conditions for various fluxes (in cm^{-2}·sec^{-1}) at the upper boundary.

Figure 2.5 shows a number of model calculations for nighttime stationary conditions in the $F2$-region, for various fluxes which are assumed at the upper boundary. Even though the stationary conception is not a suitable description of real nighttime $F2$-region (cf. para. 4.2), while under nonstationary conditions rapid flux variations at the upper boundary result in a changing altitude of the layer maximum /28/, a discussion of simplified idealized conditions contributes to our understanding of the mechanism of $F2$-region formation.

King and Kohl /200/ in 1965 were the first to point out the possible effects of thermospheric winds in the $F2$-region. Their study was followed by a large number of investigations on the same subject /75, 115, 130, 131, 202, 203, 242–244, 247, 272, 274, 292/. The principal effect of vertical drifts due to winds and to electric fields is a shift in the altitude of the layer maximum. The upward drift has the effect of raising the layer, while a downward drift depresses it. A vertical upward motion of the layer is accompanied by an increased electron concentration in the layer maximum, while a lowering of the layer altitude leads to a decrease in concentration. Figure 2.6 shows the results of the calculations of profiles obtained for different vertical drift velocities.

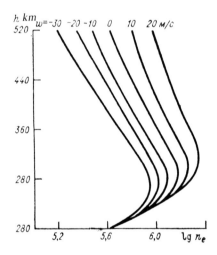

Fig. 2.6. Distributions of electron concentration for different vertical drift velocities w.

It is widely believed that the increase in n_e^m with increasing altitude of the layer maximum is linked to a slower recombination rate at higher altitudes. This is also the explanation given for the decrease in n_e^m as the layer moves downwards. However, it is readily shown that a simple mechanical shift of the layer, upwards or downwards, would not appreciably alter the electron concentration n_e^m in the layer.

In fact, if the variations in the altitude are not too large, we may write $[O] = [O]_0 \exp(-h/H)$ and $\beta = \beta_0 \exp(-1.75h/H)$ where $[O]_0$ and β_0 are the concentra-tion of atomic oxygen and coefficient of linear recombination at some level h_0, respectively. Let the layer change its altitude by Δh, which is equivalent to shifting the reading level by the same amount (Δh). For this altitude we may write

$$\Delta \ln [O] = -\Delta h/H \text{ and } \Delta \ln \beta = -1.75\Delta h/H$$

i.e.,

$$\Delta \lg \beta = 1.75\Delta \lg [O]$$

Using the relationship (2.45) at $\Delta w = 0$:

$$\Delta \lg n_e^m = 1.08\Delta \lg [O] - 0.65\Delta \lg \beta$$

This expression shows that the change of $\lg n_e^m$ caused by the change in $\lg [O]$ is almost fully cancelled out by the change in $\lg \beta$, and that the resulting $\Delta \lg n_e^m$ is close to zero due to $1.75 \times 0.65 = 1.14 \approx 1$. The physical meaning of this result is that as the altitude of the layer maximum increases, the decrease in the recombination rate is proportional to β, but the photoionization rate q also decreases proportionally to $[O]$, and that these two effects almost completely cancel each other, owing to the difference in scale heights and the different values of pre-log $[O]$ and pre-log β coefficients in expression (2.45).

It follows that the changes in the electron concentration n_e^m produced by the vertical drift should not be related to the change in the relative rates of ionization and recombination processes, but to the plasma transfer. Thus, the upward drift reduces the downward rate of diffusion, and prevents the plasma from streaming into the region of enhanced recombination, with consequent increase in electron concentration. The downward drift is superposed on the diffusion rate, accelerating the flow of the plasma into the region of intensive recombination and thus the value of n_e^m decreases. It follows from the equations (2.45) and (2.46) that the vertical drift significantly affects both the altitude of the layer maximum h_m and n_e^m. A 10 m/sec change in the drift rate results in a change of 0.09 in the value of $\lg n_e^m$ and in a 15 km change in the altitude h_m.

Calculations /78, 103, 109, 110, 161, 162, 202, 245, 248, 273/ and results of observations made by the method of incoherent scatter /79, 80, 140, 174, 257, 258/ show that the direction of the wind is mainly polewards by day, and towards the equator by night. As a result, the layer maximum shifts downwards during the day; however, at night it is at a higher level than it would have been in the absence of drift. The significance of this effect is crucial in interpreting the parameter variations of the layer maximum. Thus, the diurnal variations of the altitude of this maximum, which may reach 100 km or even more, cannot be explained without reference to thermospheric winds. The problem of formation of the nighttime $F2$-region can be easily explained if winds and observations of plasma fluxes from the protonosphere are taken into account. This problem will be discussed in more detail in sec. 4.2.

Thus, thermospheric winds are an important physical factor, which must be allowed for in calculating the $F2$-region. However, as already seen, such calculations involve major difficulties.

2.3.3 Variation of n_e with the Plasma Temperature. The temperatures T_e and T_i determine a large number of parameters which describe the distribution of electron concentration in the $F2$-region. These include the coefficient of diffusion, rate constants of dissociative recombination, and the plasma scale height. We must identify the problems whose solution require the knowledge of the values of T_e and T_i.

According to the results of the observations made in the $F2$-region /99, 116, 141, 146, 270, 282/ the electron concentration is inversely proportional to T_e, due to the specific mechanism of local cooling of the electron gas. A related effect is the inversion in the distribution of T_e with the altitude /144–146/ – the profile of the electron temperature displays a trough at high electron concentrations (of the order of 10^6 cm^{-3}) above the altitude of the layer maximum. This effect is mainly observed in winter when the n_e^m-values are high.

Below the layer maximum the relative abundance of the molecular ions O_2^+ and NO^+ increases. Thus, according to /257/, at 200 km altitude the $[O^+]/n_e$ ratio is 80% in winter and 65% in summer. Molecular ions are lost in a dissociative recombination process, the rate constant being inversely proportional to T_e (cf. para. 2.1.1). Therefore, in the computation of electron concentration in the $F1$-region, where molecular ions predominate, the electron temperature which may be considerably higher than that of the neutral particles in daytime, should be taken into account /145, 146/.

At higher altitudes, above the layer maximum, under conditions close to diffusional-gravitational equilibrium, which is in fact the case for the most part of the day (cf. para. 2.3.2), the distribution of electron concentration with the altitude is determined by the scale height of the plasma, which depends on the values of T_e and T_i. With increase of the plasma temperature, the scale height of the electron concentration above the layer maximum increases. However, the parameters of the maximum itself are not very sensitive to changes in T_i and T_e, as has already been noted /274/. The same conclusion can also be reached by analyzing the expression for the plasma flux

$$\phi = -D_a \left[\frac{\partial n_e}{\partial h} + n_e \left(\frac{m_i g}{k(T_e + T_i)} + \frac{\partial (T_e + T_i)}{dh} \frac{1}{T_e + T_i} \right) \right]$$

which yields the following expression, for the region of the maximum, $D_a = b(T_e + T_i)/\nu(O^+, n)$:

$$\phi_m = - \frac{b}{\nu(O^+, n)} n_e^m \left\{ m_i g + \frac{\partial}{\partial h} (T_e + T_i) \right\}$$

It is apparent from this expression that the flux in which the layer maximum is formed depends only on the temperature gradient, and not on the values of T_e and T_i themselves. There is some residual dependence on the ion temperature T_i which is contained in the diffusion coefficient by way of the collision frequency, $\nu(O^+, n)$. The term involving the temperature gradient $\frac{\partial}{\partial h}(T_e + T_e)$, usually contributes 15–20% to the value of H_p.

This result shows that for solving problems that only involve the calculation of layer maximum parameters, approximate values for the plasma temperature may be used.

CHAPTER 3. CHOOSING AERONOMIC PARAMETERS

Having formulated the equations describing all the basic physical processes involving the ions in the $F2$-region, we should now specify precisely the conditions of our problem. This means that definite values must be assigned to the constants in the equations, to the boundary conditions and to aeronomic parameters, taking into account their dependence on the geo-heliophysical, physical and physiochemical conditions in the ionosphere at various locations and times.

The principal parameters are: the intensity and the spectrum of the solar radiation, temperature and concentration of particles of the neutral atmosphere, and vertical drifts. It is also important to know the values of the electric fields, heat fluxes, fluxes of charged particles at the upper boundary of the ionosphere etc. It is important to specify the rate constants of various physical processes involved in the equations, such as ionization, recombination and diffusion, ion-molecule reactions, excitation, quenching etc.

It is apparent that the number of the parameters is very large, and since these parameters vary with time, coordinates, altitude, and with helio-geophysical conditions (i.e., solar and geomagnetic activities, the season of the year etc.), the complexity of the task as a whole is clear. This difficulty is further compounded by the fact that many different values of these parameters have been postulated, and that the nature of the their variation has not yet been finally established.

In view of the points mentioned above, it is rather difficult to develop an effective approach to a general solution of this complex problem. However, if we restrict our task to morphological studies alone, then clearly the parameters can be placed in order of importance and the most important one treated first. In this way the fundamental physical mechanisms accounting for the principal features of behavior of the $F2$-region can be determined and isolated.

Internal compatibility of the parameter system is also very important. However, compatibility alone is not enough. The number of parameters is large and, when they are inserted in equations, they frequently cancel one another. For this reason correct computation results may in general be obtained for several different combinations of suitably selected parameter values. In practice, it is difficult to reach an unambiguous solution and there are as many solutions proposed as there are different workers. The solutions based on the most complete and accurate experimental data should be considered as the more significant. However, parameter values corresponding to the most realistic solutions in physical terms are considered to be the most significant ones.

We shall first review the available data for each parameter, after which we will confirm our ultimate choice by comparison with ionospheric data. The choice of the parameters will be based, in the first instance, on direct experimental values. If only few such values are available, we shall use the numerous indirect data, obtained by observations of the upper atmosphere and ionosphere. The drawback of these data is that they depend on a large number of parameters which are difficult to compensate for, but if this is performed satisfactorily, the conditions obtained most closely resemble these in the ionosphere. In particular, our selection of the aeronomic parameters will also be based on the results of studies of short-wave solar radiation and its effect on the ionic composition at altitudes between 140 and 200 km, and on the E-region of the ionosphere.

3.1 SHORT-WAVE SOLAR RADIATION

For short-wave solar radiation, the most important factor is the relative variation of the radiation intensity, I_o, with solar activity, which will be considered first. After this the absolute I_o-value will be discussed.

The ability to estimate relative changes of ultraviolet solar radiation is important in theoretical ionospheric investigations, since it enables variations to be taken into account. The results of short-wave solar radiation determinations in various spectral ranges, performed prior to 1965, reviewed critically, showed /16/ that in each spectral interval the radiation intensity regularly increases with the increase of the solar activity. The rate of increase of radiation intensity increases with the ionization potential of the ion giving rise to the radiation.

This finding was used to plot the complete spectrum of short-wave radiation for the "minimum" ($F_{10.7} = 80$) and the "maximum" ($F_{10.7} = 200 - 250$) solar activities. The spectra obtained by Ivanov-Kholodnyi /16/ will be found in /33/. The total flux of the ionizing radiation in the $0.1 - 102.7$ nm spectral interval varies between 2.5 mW/m^2 at the minimum and 7.5 mW/m^2 at the maximum of the solar cycle. This three-fold increase at the maximum, as compared to the minimum of the solar activity, is caused by the fact that the variations in the intensity of the radiation emitted by the transition zone in the solar atmosphere (which supplies the bulk of the energy in the short-wave region of the spectrum) are caused by the intensity changes in radiation emitted by the solar corona and the chromosphere. Numerous observations made during solar eclipses indicate that the radiation intensity of the hotter corona in the spectral lines increases by a factor of $5 - 10$ between the minimum and the maximum, while the corresponding factor is only $1.3 - 1.5$ for the radiation intensity of the chromosphere, which is cooler than the transition zone.

Since the solar activity varies not only during the solar cycle, but also from day to day, the radiation may likewise be expected to vary from day to day. The effect of solar activity may be estimated from the radio-wave solar radiation in the decimeter range (e.g., at the wavelength of 10.7 cm), which originates from the transition region, as does the short-wave radiation. However, this conclusion is not generally accepted, and will therefore be considered at some length below.

Hall et al. noted in 1969 /170/ that UV radiation only varied regularly with the solar cycle, while the day-to-day changes in the radiation intensities of individual lines were found by rocket measurements not to have a definite dependence on the $F_{10.7}$ index. Figure 3.1 shows the results of 11 determinations of I_o' for the $\lambda = 30.4$ nm line /170/ made by rocket-borne instruments; the intensities are given in relative units. Unit intensity was taken as the intensity measured in a rocket /179/ on 11th March, 1967, at $F_{10.7} = 144$. The values show considerable scatter. In the $F_{10.7} \approx 130$ range the intensity varies roughly by a factor or two.

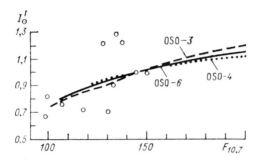

Fig. 3.1. Variation of the relative intensity I_o' of the $\lambda = 30.4$ nm line in the solar spectrum as a function of the index $F_{10.7}$ from the data furnished by rocket-borne instruments (circles) and instruments carried by OSO-type satellites /19/.

The results of Hall et al. /170/ cast doubt on the feasibility of calculating the short-wave radiation flux I_o from solar activity indexes. The feasibility of theoretical computations of the $F2$-region and other ionospheric regions was questionable until systematic observations of short-wave ultraviolet solar radiation began.

On the other hand, it was shown by Ivanov-Kholodnyi /19/, who adopted the approach in /16/, that in principle it is possible to forecast the relative day-to-day variations in the intensity of short-wave solar radiation. This conclusion is essentially based on relative determinations of I_o, since despite the large scatter in the data obtained by rocket measurements, there is satisfactory *average* correlation between the line intensities and the index $F_{10.7}$, as shown by many results of systematic satellite measurements.

The average I_o as a function of $F_{10.7}$, of to the data obtained by three OSO satellites /171, 179, 286, 300/, are shown in Fig. 3.1. It is apparent that in the $F_{10.7}$ $=100 - 200$ range they practically coincide, the maximum deviation at the boundary of the region being $6 - 7\%$. It should be noted that the agreement of relative measurements is implied; the absolute intensity values at $\lambda = 30.4$ nm, as obtained, for example, by /286, 300/, are about 1.5 times higher than those obtained in /179, 171/.

But what about the rocket measurements? It was found /19/ that the variations of line intensities depend both on the solar activity and on the ionization potential. This is indicated by results obtained on board satellites and in rocket carriers.

It was shown that the ratios between pairs of respective lines in individual rocket experiments agree with one another. The data of satellite experiments /171/, plotted on a logarithmic scale, only show a $0.02 - 0.05$ deviation from the average curve describing the variation of the relative intensities of spectral lines with the ionization potential. It follows that the scatter of the absolute intensities in different rocket experiments is due to the inhomogeneity of the data, i.e., to the different absolute calibrations of the intensity in individual experiments.

In order that the relationship between the flux of short-wave solar radiation and $F_{10.7}$ might be effectively utilized, we must know its degree of accuracy. It was shown /19/ that, as compared with the average variation of the relative intensities with $F_{10.7}$, obtained from the satellite data, the deviations of the relative data obtained on board rockets do not exceed ±0.05 on a logarithmic scale (±12%) in most (about 90%) cases. The results obtained in /19/ are scanty, but the author's conclusions have been confirmed by other workers.

A large series of systematic measurements of the intensity of the λ = 30.4 nm line was performed on board satellite OSO–4 /286/. The accuracy of the absolute determinations was ±30%, while the relative accuracy was probably ±5%. The experimental work covered the periods October–November 1967, and June 1968 – December 1969, i.e., more than 250 days. A comparison of the variations of intensity, I (λ = 30.4 nm) from day to day with $F_{10.7}$ index is shown in Fig. 3.2 /286/. The linear correlation ratio may be written in the form:

$$I(\lambda = 30.4 \text{ nm}) = 7.82 + 0.0097F_{10.7}, \qquad (3.1)$$

where the unit of $I\lambda$ = 30.4 nm) is 10^9 photons/(sec·cm^2).

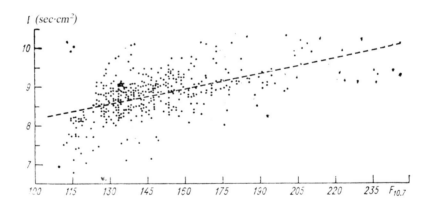

Fig. 3.2. Variation of the intensity of the line λ = 30.4 nm with the index $F_{10.7}$ from the results obtained on board the satellite OSO–4 /286/.
The dotted line represents the linear relationship (3.1) which does not reflect the effect of a faster decrease in intensity at low $F_{10.7}$ values.

The mean square error, σ, of the value of I is 0.463 (or about 5%) /286/. It is apparent from Fig. 3.2 that the deviations from the mean may be larger than 3σ in individual cases and may reach 20–25% (about ±0.1 on the logarithmic scale). These results are in agreement with the error estimates given above and with other radiation lines obtained from the rocket data /19/.

Obviously, the total flux of short-wave solar radiation, the relative value of which varies with $F_{10.7}$ in a manner approximately similar to that of the line $\lambda = 30.4$ nm, would be expected to vary in a similar manner.

Thus, the error involved in the estimates of the flux of short-wave solar radiation from $F_{10.7}$ should not exceed $0.02 - 0.05$ on the logarithmic scale (5–12%) in 90% of the cases, though it may be as large as 20–25% in individual cases. Such accuracy is sufficient for solving many aeronomic problems.

We may ask why the value of $F_{10.7}$ does not represent an unambiguous description of the UV flux. This is because variations of the solar activity are caused by changes in the number and the brightness of active solar regions. The fluxes of short-wave solar radiation and $F_{10.7}$ vary in the same manner with the overall surface area of the active zones, but in a somewhat differing manner with brightness (i.e., the temperature composition) and location of these zones on the disk of the Sun. Hence differences between both radiation fluxes occur.

Recently doubts arose again with respect to the suitability of the $F_{10.7}$ index as a quantitative characteristic of the UV flux /180, 261, 263/, on account of the unsatisfactory correlation between the day-to-day variations of the two parameters. It was pointed out /181/ that during the $F_{10.7}$minimum period a local maximum of the intensity of the HeI 58.4 nm line was observed. Statistical analysis of measurements carried out on board satellite AEROS-A during 8 months in 1973 /236/ indicated that none of the classical indexes of solar activity represents an accurate description of day-to-day UV flux, which is usually intermediate between the variation of $F_{10.7}$ and the sunspot number, W. Heroux and Hinterlagger /178, 181/ even found qualitative differences in the evolution of short-wave solar radiation between the 20th and the 21st cycles of solar activity. While during the 20th cycle the 30–122 nm radiation flux was practically constant (with a variation of up to 10% in the $F_{10.7}$ range between 77 and 177), the variation amplitudes of the two parameters became equal during the 21st cycle*. The radiation intensity emitted by the quiet solar disk in the 21st cycle proved more intensive than in the 20th. However, all these findings are still to be confirmed, and for this purpose effective, permanent, direct monitoring of short-wave solar flux is required.

Such measurements have not yet been carried out, and the evaluation of the relative variations of the total short-wave solar fulx from the $F_{10.7}$ index remains an important, and possibly the only practical method. We have said that the error involved in the method is 20–25% at most; in fact, the errors of all the experimental values quoted in /178, 181, 236, 261/ remained within these limits, even when the maximum intensity of the radiation and the minimum $F_{10.7}$ value (and *vice versa*) coincided. If we consider intensities of individual lines rather than that of the total flux, poor correlation with $F_{10.7}$ may be obtained /178, 180/. However, these effects may be ignored in calculating the F2-region of the atmosphere, in which only the total short-wave radiation flux of the Sun is relevant.

Since the index $F_{10.7}$ is an approximate (to within 10–20%) measure of the flux of short-wave solar radiation, direct satellite results obtained for specific days, rather

---

*However, according to /261/, in 1973 alone, i.e., during the 20th cycle, the average variation factor in the radiation flux between 15 and 103 nm was 1.5, when $F_{10.7}$ changed from 150 to 120.

than average estimates, are used in certain aeronomic studies in which high accuracy is required. Thus, follow-up determinations of X-radiations of the magnitude of a few per cent were used to analyse the effects of flares in the E-region /26, 27/, while satellite measurements of λ = 30.4 nm line radiation were employed /34/ in the analysis of day-to-day "fine" variations in the E-region. In other cases, when an estimate of the short-wave solar radiation is sufficient, its correlation with the index $F_{10.7}$, discussed above, is employed.

To calculate the ionization of the upper atmosphere, data are required on the total short-wave radiation spectrum throughout the spectral range, extending from a few tenths of a nanometer to the wavelength of 103.7 nm, which is the ionization threshold of atmospheric components. Ivanov-Kholodnyi and Firsov /25/ calculated the total spectrum of this radiation at four definite solar activity levels. During the calculation of variations of about 500 spectral radiation lines and of the coninuum, allowance was made for the ionization potential of the radiating ion or atom, and the average dependence on $F_{10.7}$ was taken in accordance with /19/.

Since the reference value taken in /25/ was the solar activity spectrum for $F_{10.7}$ = 144 as published in /179/, with its intensity reduced approximately by a factor of 2 (see below), the intensity values reported in Table 2.1 for 150 spectral intervals are also reduced by this factor. Table 3.1 shows the data of energy distribution over 12 large spectral intervals, corrected for this factor, since a finer breakdown of the spectrum is not required to calculate the ion formation rate in the ionospheric F2-region. Figure 3.3 shows the variation of the total flux as a function of $F_{10.7}$. It is apparent that the total energy of solar radiation is doubled at $F_{10.7}$= 200 as compared to $F_{10.7}$ = 100. Thus, the variations of the solar activity from the minimum ($F_{10.7} \sim 70$) to the maximum value ($F_{10.7} \sim 250$), correspond to an increase by about a factor of 4. A similar overall variation is displayed in the intensity of the bulk portion of the radiation, between 20 and 63 nm, though the bright HeII line (30.4 nm) varies at a somewhat reduced rate. Longer-wave radiation varies even more slowly (roughly in proportion to the square root of the variation factor of $F_{10.7}$), while in shorter wave length portion of the spectrum ($\lambda \leq 20$ nm), which contains about one-third of the total energy, it displays a considerable increase with decreasing wavelength.

TABLE 3.1

Distribution of short wave solar radiation flux over 12 main spectral intervals, $\Delta\lambda$ for four $F_{10.7}$-values.

$\Delta\lambda$ nm	$F_{10.7}$				$\Delta\lambda$ nm	$F_{10.7}$			
	100	120	144	200		100	120	144	200
103.7–91.1	0.458	0.544	0.630	0.736	20.5–18.0	0.326	0.434	0.552	0.713
91.1–80.0	0.308	0.366	0.412	0.468	18.0–16.5	0.182	0.236	0.298	0.380
80.0–63.0	0.080	0.098	0.126	0.144	16.5–12.0	0.056	0.780	0.104	0.138
63.0–46.0	0.212	0.270	0.344	0.420	12.0–6.0	0.128	0.170	0.212	0.264
46.0–31.0	0.260	0.364	0.480	0.650	6.0–0.1	0.142	0.190	0.242	0.364
31.0–28.0	0.582	0.730	0.890	1.106	103.7–0.1	3.09	3.964	4.920	6.226
28.0–20.5	0.356	0.484	0.624	0.830					

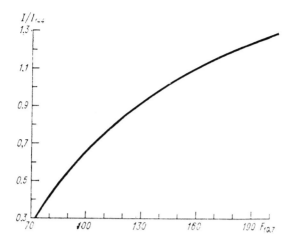

Fig. 3.3. Relative variation of the total flux of short-wave
radiation as a function of the index $F_{10.7}$ /25/. The flux at
$F_{10.7} = 144$ was chosen as reference unit.

It is apparent from Fig. 3.3 that at low $F_{10.7}$-values, the short-wave radiation flux decreases more rapidly than at larger $F_{10.7}$-values. In Fig. 3.2 the experimental points are located below the straight line corresponding to the linear relationship (3.1) even for low $F_{10.7}$-values. It must be concluded, accordingly, that the relationship between I_o and $F_{10.7}$ does not approximate a straight line throughout the range of $F_{10.7}$-values. The increase of I_o with $F_{10.7}$ is much steeper at low than at medium and high $F_{10.7}$-values.

Let us consider the absolute UV flux. As we noted previously the 1969–1970 review of calibration of reference spectrum /171, 179/ resulted in an approximate halving of the total energy as compared to the formerly quoted value of 2.5 mW/m^2 at minimum activity. However, this calibration was recently reviewed once again, and the results, which in this case gave a higher value, will now be discussed.

In a series of studies, Schmidtke /261–264/ introduced a more correct system of absolute intensity calibration with the aid of satellite measurements, and concluded that the short-wave solar radiation intensity, as reported by Hinteregger, was too low by a factor of 2–3. This important conclusion, published in 1976–1977, was confirmed by the reports of the Special Session of Working Group IV of COSPAR, by comparing all available measurements and allowing for the variation of the short-wave solar radiation flux with time, with careful analysis of the experimental errors /122, 123/.

According to the data obtained by Schmidtke on 19th January 1973, the short-wave solar flux at $F_{10.7} = 95$ between 16 and 103 nm, accounting for about 90% of the total radiation energy, was 3.37 mW/m^2 /122, 123, 261/. This is about 1.5 times higher than the value obtained by Hinteregger /179/, and a similar result was obtained for higher activity level at $F_{10.7} = 144$, when the expected flux was 1.7 times higher than at $F_{10.7} = 95$. Thus, original estimates of the total UV /16, 33/ –

2.5 and 7.5 mW/m^2 at $F_{10.7}$-values of 80 and 200 respectively* – have been confirmed by new experimental data on variations of the short-wave flux with $F_{10.7}$, taken 10 years after the original data were obtained.

Our own original theoretical estimates of the UV flux as a function of the solar activity /32/, which were determined in 1963 /12, 60/ at 5–10 mW/m^2, and modified in 1966–1969 /16,33/ to 7.5 mW/m^2 at maximum solar activity ($F_{10.7} = 200 - 250$), are in agreement with the most recent results. It should also be noted that Schmidtke's new spectrum /122, 123, 261/ becomes closely similar to that published /36/ for $F_{10.7} = 100$, if its intensity is multiplied by a factor of 2–3 (except for the 91.1 – 102.7 nm interval, where the value of this factor is 1.5 or, more accurately, increases uniformly from 1.5 to 3.5 in the spectral interval between 100 and 16.5 nm). According to a number of other studies, too, Hinteregger's data /179/ are low by a factor of 2–3. These indirect conclusions depend to a large extent on the particular choice of the system of aeronomic parameters (reaction constants, atmospheric models etc.), which in itself requires verification. Accordingly, they are not nearly as reliable as the results of direct, well organized mesaurements.

3.2 PARAMETERS OF NEUTRAL ATMOSPHERE

Due to the effect of short-wave solar radiation, the upper atmosphere is partially ionized. The ionized part of the atmosphere constitutes the ionosphere. The various changes occurring in the neutral part of the atmosphere affect the ionosphere to varying extents. Consequently, the changes occurring in the ionosphere cannot be understood unless the variations iñ the parameters of the neutral atmosphere are known.

The principal parameters of the neutral atmosphere are its temperature, T, and the concentrations of its components (N_2, O_2, O, He, H, Ar). These parameters may be used in calculating the density ρ, pressure p, scale height H, the relative concentrations, $[O]/[N_2]$, $[O_2]/[N_2]$, etc. At high altitudes in the exosphere (above 300 km) the temperature, T_∞, becomes constant.

All atmospheric parameters vary with the helio-geophysical conditions. These include the geometric parameters: the geographic coordinates φ and λ, the altitude h and also the time, including the serial number of the day in the year, d, and the hour, t.

It is also important to take into account the solar activity, I_o, or $F_{10.7}$, and the geomagnetic activity A_p or K_p during previous time periods. We have ten parameters in all: φ, λ, h, t, I_o, $F_{10.7}$, $\bar{F}_{10.7}$, A_p, K_p.

Information about the neutral atmosphere is obtained by various direct and indirect methods – the basic information by direct measurements made on board rockets and satellites and in incoherent scatter installations. We shall give a brief account of the results of such studies, which are of interest in calculations of the $F2$-region of the ionosphere.

---

* The value of 7.5mW/cm^2 at $F_{10.7} = 200$, and the conclusion to the effect that UV flux decreases 2.6 times at $F_{10.7} = 70$, are due to Chernyshev /72/, who employed a similar approach and used recent experimental data.

Since it is clearly impossible to describe all the possible variations of the numerous parameters of the upper atmosphere, we shall restrict our considerations to a small number of parameters relevant to the material to be considered in the chapters that follow. For more detailed information the reader is referred to special monographs and surveys /39, 42, 50, 70/ (cf. also /17, 18, 20, 33/). We shall consider the following three important problems: annual variations in the temperature and composition of the atmosphere; variation of these parameters with the solar activity; and their planetary distribution.

3.2.1 Annual Variations. Annual variations of the temperature are usually described as those between winter and summer (seasonal variations) and from solstice to equinox (semiannual variations). In the former case the variation extends over the entire year, in the latter over half a year.

A typical example of seasonal variations is the variation of the exospheric temperature T_∞, which is lowest in winter and highest in summer. The values of T_∞ and their summer-to-winter gradient are a function of the latitude, solar activity and time of the day. Thus, for instance, modern satellite measurements (observation of the Doppler width of the emitted lines) indicate that at $F_{10.7} = 150$ at moderate latitudes, $T_\infty \approx 950°K$ in winter and $T_\infty \approx 1120°K$ in summer at the poles T_∞ vary by 800–1200°K during the year /285/. Formerly seasonal changes of T_∞ were believed to be larger.

Seasonal variations of the composition of the atmosphere are more complex. An extensive program of rocket and satellite studies was carried out, and the results obtained were integrated into a coherent system. Even the concentrations of the principal components of the atmosphere undergo complex variations. Thus, for instance, at altitudes above 200 km the N_2 concentration is higher in summer than in winter, owing to the seasonal variations of T_∞. At altitudes of about 100 km, on the other hand, the winter concentrations of molecular nitrogen are 1.3 – 1.5 times higher than in summer /169/. At altitudes of 90 km and 150 to 200 km there is an isopycnic level, i.e., a level at which the concentration of N_2 remains practically unchanged throughout the year. Thus, in describing the variations in $[N_2]$, it is necessary not only to allow for T_∞ -variations during the year, but also for the seasonal variations of the molecules at the lower boundary, near $h = 100$ km (in earlier atmospheric models it was assumed, although the evidence was insufficient, that the temperature and the concentration of the molecules remain unchanged at the altitude of 120 km).

Atomic oxygen was also observed to display typical semiannual variations. A major semiannual component of the variation of the relative concentration of atomic oxygen at altitudes of 120–130 km was detected by rocket measurements /38, 119/. However, a more convincing argument in favor of semiannual variations of the absolute of atomic oxygen concentration is the fact that the variations of the overall density of the atmosphere ρ at 300–800 km altitudes, where atomic oxygen constitutes the main component, have a semiannual component /18/. It should be noted that semiannual variations of the overall density of the atmosphere ρ were detected from the drag on the first Soviet satellites as early as 1960. We now know that this property of the upper atmosphere is global, but until very recently, no physical explanation for it could be given.

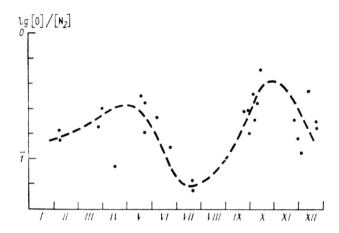

Fig. 3.4. Annual variations of the concentration ratio [O]/[N₂] from
the results of rocket measurements at the altitude of 130 km /2/.

Figure 3.4 shows the annual variations of the concentration ratio $[O]/[N_2]$,
obtained from rocket measurements /2/. For other similar results, see /38, 40, 119/. It
is important to note the following features: 1) the semiannual maxima are delayed by
about one month with respect to the equinox; 2) the autumn peak is somewhat
higher than the spring peak (about 1.5 times); and 3) the [O] concentration increases
about three times* during the period from summer to the equinox, and 1½ – 2 times
between summer and winter (this means that, in addition to the semiannual
component, there is also a weaker seasonal component of [O] variation). Similar
conclusions can be reached on the basis of satellite data on the overall density of the
atmosphere at 300–800 km latitudes /18/, but according to these data the amplitude
of semiannual and seasonal variations is usually smaller than ₒthat observed with
rocket-borne instruments; moreover, it was noted that this amplitude varied from
year to year. Ivanov-Kholodnyi and Katyushina /24/ proposed one of the first
atmospheric models for altitudes of 130–200 km, which reflects the average semiannual variations of the composition and the overall density of the atmosphere. In
order to understand the reasons for the semiannual variations of atomic oxygen, it is
necessary to consider its mechanism of formation in the upper atmosphere.

Atomic oxygen is a special component of the upper atmosphere in the sense that,
unlike molecular nitrogen or argon, for example, it does not enter from below, but is
formed in the upper atmosphere itself by dissociation of O_2 under the effect of solar
radiation. The rate of this dissociation, which is proportional to the concentration of
molecular oxygen and to the intensity of the ionizing radiation, increases with
decreasing altitude, reaching its maximum value at the altitude of about 80 km. The
maximum concentration of atomic oxygen is formed at a higher altitude – about

---

* According to the most recent measurements by rocket-borne instruments, carried out in the zone
of maximum atomic oxygen concentration (h = 90–95 km), the latter increases by a factor of 2 –3
between the summer and the equinox /20/.

90–95 km. It is important to bear in mind that the atomic oxygen in the upper atmosphere (about 90–100 km) is in diffusional-gravitational and not in photochemical equilibrium /23, 59/.

Thus, the conditions resemble those causing the $F2$-layer formation; the maximum concentration n_c in the layer does not correspond to the maximum rate of ion formation, due to the effect of diffusion. As in the case of the $F2$-layer, the maximum atomic oxygen concentration is encountered in the altitude range (90–95 km) at which equilibrium prevails between the effective rate of vertical transport by turbulent diffusion and the effective time of the recombination process.

Since the effectiveness of turbulent diffusion varies with the solar irradiation /196/, it undergoes considerable annual variations, being at a maximum in summer. Thus, the oxygen atom outflow downwards is more pronounced in summer, and thus its concentration decreases to its minimum value. However, since the diurnal dissociation of O_2 in the upper atmosphere is larger during the summer than in winter, the observed [O] concentration is the result of a dynamic equilibrium between the two opposing processes, diffusion and dissociation. It was shown by Antonova and Katyushina /2/ that this equilibrium results in maximum [O] concentrations approximately at the equinoxes.

The mechanism proposed in /2/ for the considerable increase in the turbulent transport coefficient during the summer is noteworthy, explaining as it does not only the semiannual variations of [O], but also giving a quantitative description of the effects of summertime decrease in the concentration of the light helium atoms /269/, accompanied by an increase in the concentrations of the heavy argon atoms /2/ and O_2 molecules /36/. Since the lifetime of the atomic oxygen layer is about one month, the dynamic variety of the model also yields the corresponding time late in the appearance of [O] peaks with respect to the equinox.

Another explanation of the seasonal changes in the composition of the atmosphere, suggested by /137–139, 143/, relates these changes to the seasonal changes in the system of thermospheric winds, which may transport atomic oxygen from the summer to the winter hemisphere, thus altering the ratio between [O] and [O_2] + [N_2].

Finally, we may consider the variations of [O_2] during the course of the year. Determinations made by the incoherent scatter method /77, 268/, by observations of airglow at night /227/, and by determining the absorption of UV radiation /212/ revealed the presence of seasonal changes in the concentration of molecular oxygen at altitudes above 100–150 km, the values in summer being much higher (3–6 times) than those in winter. The signs of those changes are opposite to the seasonal concentration changes of molecular nitrogen and atomic oxygen at about 100 km altitude. Such changes in [O_2], together with semiannual changes in [O], are essential in the theory of ionosphere formation, in order to explain the observed seasonal variations in the E-region /26/ and the ion composition at altitudes of 130–200 km /1/. Possible reasons for the increase in the concentration of molecular oxygen were discussed above.

Other atmospheric components are less relevant to analysis of the ionospheric $F2$-layer.

3.2.2 Variation with Solar Activity. The variations of the upper atmosphere parameters as a result of variations in the solar activity are principally determined by the relationship between the temperature T_∞ and the solar activity level.

As a result of prolonged measurements of the drag effects on artificial Earth satellites it was found that the changes in T_∞, both at midday and around midnight, are closely connected with the changes in the index $F_{10.7}$. The computation techniques of the exact effect of solar activity have gradually improved. Since this problem is very important, we shall consider it in detail here.

Figure 3.5 shows that $F_{10.7}$ varies linearly with the surface area of sunspots. The slopes of the averaged graphs are the same for different years. However, during years with a high solar activity, the $F_{10.7}$-value corresponding to a particular S_W is higher than in years with a low activity. The systematic variation of the instantaneous $F_{10.7}$-value at a particular S_W-value may be allowed for by using a special technique – the introduction of an index of the average solar activity $\bar{F}_{10.7}$ (the $F_{10.7}$-value, averaged over several 27-day periods).

Figure 3.6 shows F_o as a function of $\bar{F}$ (F_o are the $F_{10.7}$-values obtained on extrapolating the graphs in Fig. 3.5 to $S_W = O$). Each point on the graph is the result obtained during one 27-day period of rotation of the Sun /191/. A similar systematic variation of the exospheric temperature T_∞ as a function of the phase of the solar cycle is allowed for in atmospheric models in the same manner – by the introduction of $\bar{F}$.

Some modifications have been introduced in the course of time both in the formulas describing the dependence of T_∞ on F and $\bar{F}$, and by averaging F. Formerly, the $\bar{F}$-value was taken as the average F-value over three rotations of the Sun /188/, when the following formula was utilized:

$$T_\infty^{min} = 379 + 3.24\bar{F} + 1.3(F - \bar{F}). \tag{3.2}$$

Subsequently /192/, F was averaged over six rotations of the Sun, and a formula containing second-power terms was used:

$$T_\infty = 350.9 + 5.163\bar{F} + 1.954(F - \bar{F}) - 0.00492\bar{F}^2 -$$
$$-0.0783(F - \bar{F}(^2. \tag{3.3}$$

A similar formula is employed in recently proposed atmospheric models /176, 177, 285/:

$$T_\infty = \text{const} + a(F - \bar{F}) + b\bar{F} + c(F - \bar{F})^2. \tag{3.4}$$

The values of the coefficients, a, b and c, for the various models are shown in Table 3.2.

The following comment may be made on Table 3.2: in the JACCHIA–73 model alone all coefficients were about 1½ times higher than in JACCHIA–71, while in the more recent models /176, 177, 285/ they were close to the coefficients in /188/.

It is important to note that long-term (i.e., 11-year period variations, which depend on $\bar{F}$, proved to be 2–3 times higher than short-term variations (e.g., 27-day variations),

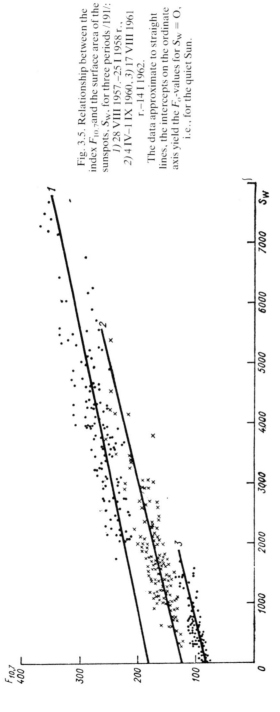

Fig. 3.5. Relationship between the index $F_{10.7}$ and the surface area of the sunspots. S_w, for three periods /191/: 1) 28 VIII 1957.–25 I 1958 r., 2) 4 IV–1 IX 1960, 3) 17 VIII 1961 r.–14 I 1962.

The data approximate to straight lines, the intercepts on the ordinate axis yield the F_0-values for $S_w = 0$, i.e., for the quiet Sun.

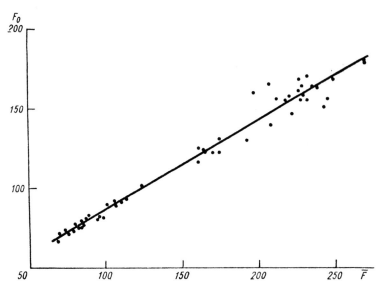

Fig. 3.6. The parameter F_o as a function of the average value of the index $\bar{F}$ /191/.

TABLE 3.2

Coefficients a, b and c, for various atmospheric models /285/.

Model	a	b	c	Reference
JACCHIA–71	1.300	3.240	0	/188/
JACCHIA–73	1.954	5.163	–0.0078	/192/
OGO–6	1.185	2.935	–0.0056	/176/
MSIS	1.247	3.280	–0.0056	/177/
M	1.325	2.531	–0.0056	/285/

which are determined by the terms of (3.2) giving T_∞ as a function of $(F - \bar{F})$.

As has been noted, the introduction of the parameter $\bar{F}$ eliminated the systematic evolution of T_∞ as a function of $F_{10.7}$, i.e., made it possible to calculate the T_∞-values more accurately under specified helio-geophysical conditions. At the same time it became inherently impossible to make a prediction of the T_∞-value or to predict the atmospheric model, since $\bar{F}$ is determined both from the data obtained 3 months (or 1½ months) prior to a given day, and from the data for a similar period of time following that day. Any prediction method must accordingly obviate this difficulty (cf. Chap. 5).   In the JACCHIA–77 model /190/, averaging over three revolutions of the Sun was again used, but the weight function was also considered, and the formula became nonlinear:

$$T_{1/2} = 5.48 \bar{F}_{10.7}^{0.8} + 101.8 \bar{F}_{10.7}^{0.4}. \tag{3.5}$$

where $T_{1/2}$, is the average of the planetary minimum T_∞^{min} and maximum T_∞^{max}.

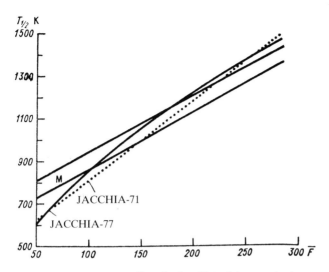

Fig. 3.7. The average diurnal value, $T_{1/2}$, of the exospheric
temperature T_∞, as a function of the average value of the index $\bar{F}$,
for four atmospheric models /176, 188, 190, 285/.
The JACCHIA–77 model /190/ is the only one to allow for the
nonlinear variation and a faster decrease of $T_{1/2}$ at small $\bar{F}$-values.

It is of interest to compare the T_∞-values obtained according to the different models
(Fig. 3.7). Such a comparison reveals that the gradients of the variation of T_∞ with $\bar{F}_{10.7}$
in the region $\bar{F}_{10.7} > 100$ are similar for all models. It should be noted that the different
models are based on data obtained by different methods: from satellite drag /188, 190/,
from mass-spectrometric measurements alone /176/ or supplemented by data obtained
by the incoherent scatter method /177/, and from the optical measurements of the
Doppler width of the oxygen red line of the airglow, taken in artificial satellites /285/.

In the new JACCHIA–77 model /190/, the T_∞-value for low $\bar{F}_{10.7}$-values is much
smaller than in other models. This accelerated decrease of T_∞ with decreasing $F_{10.7}$
during periods of weak solar activity is in agreement with the relatively more rapid
decrease of I_o under these conditions (cf. Figs. 3.2 and 3.3).

Since the accuracy of determination of T_∞ in different models is claimed as $\pm 50°K$,
the differences between individual models may be $100°K$ or more, even if $F_{10.7} > 100$
(this is also the estimated real accuracy of absolute T_∞-values).

Thus, depending on the degree of solar activity, relative changes in T_∞ are given by
atmospheric models to within $\pm 50°K$, but the estimate of the absolute value may involve
an error as large as $\pm 100°K$.

The temperature of the atmosphere also changes in the presence of atmospheric
disturbances:

$$\Delta T_\infty = k\Delta K_p, \tag{3.6}$$

where ΔK_p is the change in the geomagnetic activity index during 5–10 hours prior to the
time of the observation; $k = 25$–$50°C$ for moderate latitudes. For more detailed data
about changes in T_∞, and in the composition of the atmosphere during geomagnetic
disturbances, the reader is referred to /17, 20, 190/.

3.2.3 Planetary Distribution. With the advent of the cosmic era, the use of artificial satellites made it possible to carry out extensive systematic series of determinations of the parameters in the upper atmosphere. As a result, not only were the variations already known confirmed, but the distribution of these parameters over the globe could be established due to the recent advent of satellites with polar orbits allowing more extensive mapping of the Earth's parameters, including the polar and subpolar regions.

The global distribution (i.e., the distribution over the geographic coordinates) of upper atmosphere parameters such as ρ, T_∞, $[N^2]$ and $[O]$, may be plotted in a statistical-empirical manner, in a fashion similar to critical frequency plots f_oF2 or f_oE for the ionosphere. The starting material consists of abundant data obtained with the aid of artificial satellites, and also by ground measurements: ρ and T_∞ from the measured drag on the satellite (CIRA, JACCHIA and /95/ models); $[N_2]$ and $[O]$ by mass-spectrometric measurements from satellites /176/; T_∞ by shift satellite determination of the Doppler shifts of the 630 nm airglow line of oxygen /101/ or from measurements at the incoherent scatter stations /77/ or from both of these taken together /285/. The data obtained for each of the above parameters are subjected to spherical harmonic expansion analysis, the coefficients found being determined by statistical methods. As a result empirical models are obtained in the form of tabulated coefficients, and can be used in the calculation of the parameter distribution charts over the Earth. We may note that prior to the advent of satellites there was no way of determining the global distributions of atmospheric parameters, and there were no theoretical foundations for their calculation. We shall now discuss the principal results obtained by using empirical models.

Figures 3.8 ($h = 400$ km) and 3.9 ($h = 275$ km), taken from /95/, show the diurnal and the annual distributions of the parameters ρ, $[N_2]$, $[O]$ and $[He]$, over the geographical latitudes.

It is apparent from Fig. 3.8 that the diurnal and annual variations of ρ and $[O]$ are similar. This is due to the fact that atomic oxygen is the major species at the altitude of 400 km. In Fig. 3.9, on the other hand, at the altitude of 275 km, the diurnal latitude distribution of ρ resembles that of $[N_2]$; in particular, the maximum ρ is observed in the vicinity of the summer pole rather than in the equatorial region as in the case of $[O]$.

The plots of the annual cycles in Figs. 3.8 and 3.9 display conspicuous $[O]$ and ρ peaks, at both altitudes and at all latitudes. The latitude distribution of $[He]$ is opposite to that of $[N_2]$.

Let us consider certain characteristic features of atmospheric models. According to /95, 188/, the seasonal component of variation in ρ at the altitude of 400 km is present in the southern hemisphere only, while being pratically absent in the northern hemisphere. The ratio between the concentrations of atomic oxygen at the equinox and in summer is largest in the equatorial zone – about 2. Figures 3.8 and 3.9 also show that there are small shifts of the planetary equinoctial maximum of ρ, away from the equatorial zone towards the north in April, and towards the south in October, so that at any given northern latitude the October maximum of ρ is not necessarily larger than the April maximum. It was shown in sec. 3.2.1 that the higher October maximum follows from the analysis of the experimental data on ρ and on atomic oxygen. Such a tendency may be noted in the graphs involving $[O]$ as well.

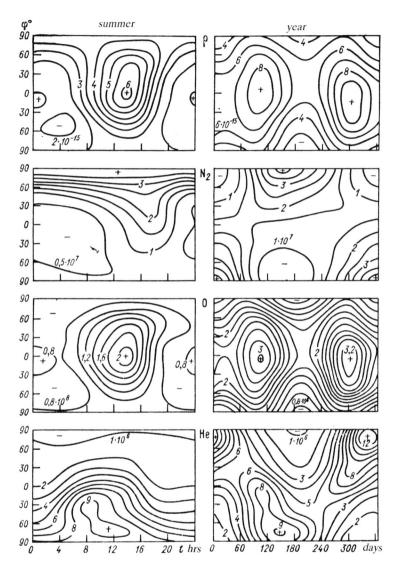

Fig. 3.8. Distribution of ρ, [N₂], [O] and [He] over the geographical latitude. Left: diurnal distribution in summer; right: annual distribution at 1500 hrs, at the altitude of 400 km, during enhanced solar activity ($F = \bar{F} = 150$) /95/.

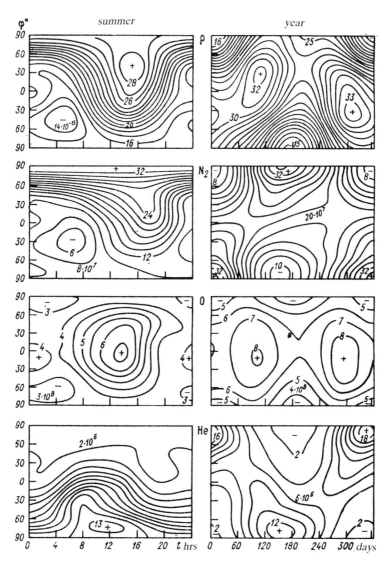

Fig. 3.9. Distribution of ρ, [N₂], [O] and [He] over the geographical latitudes.
Left: diurnal distribution in summer; right: annual distribution at 1500 hrs, at
the altitude of 275 km, during low solar activity ($F = \bar{F} = 92$) /95/.

A similar latitude distribution of various atmospheric components was formerly described in the context of OGO–6, MSIS and other models /176, 177, 297/, plotted from the *in situ* mass-spectrometric determinations conducted on satellites.

In all the empirical models listed above the distributions of atmospheric parameters differ from those in the earlier JACCHIA models /188, 192/, based on satellite drag data. This difference is most marked in the case of T_∞. According to the early JACCHIA models, the maximum T_∞ is located close to the subsolar point, while the minimum is found at the anti-solar point, whereas satellite data, based on the determination of the Doppler width of airglow and on the atmospheric composition, located these points in the vicinity of the poles. Since in JACCHIA models the input parameter is T_∞, this discrepancy also introduced other inaccuracies in the description of the global distribution of atmospheric parameters in the early models. In a later JACCHIA model /190/ the planetary extrema of the temperature during solstic periods are shifted to approximately 50° lat.

3.2.4 Analysis of and Comments on the Semi-Empirical Atmospheric Models. The upper atmosphere is a complex geophysical geophysical formation, with properties varying according to the geographical coordinates, time of the day and season of the year, and also as a function of solar and geomagnetic activities. As a result of empirical studies, the nature of these variations is now known. They now have to be integrated into a system and related to physical processes. A working hypothesis has been advanced, according to which the principal variations of the properties of the upper atmosphere with time, insofar as they are not due to diurnal and seasonal variations, are mainly caused by solar and geomagnetic activities. Semi-empirical models of the atmosphere have been established on this assumption, while the theoretical models are still unsatisfactory.

The existing semi-empirical models describe atmospheric variations to varying degrees of accuracy and completeness. Individual models also differ in their convenience for use and in the purpose they are intended to serve. Despite the recent tendency to create universal, planetary models, it is not possible to make a simplified model incorporating all the available knowledge about the upper atmosphere, particularly because certain problems have not yet been fully clarified. The limitations of any individual model may not be immediately apparent, and may only be revealed on comparing the model with the results of direct observations.

The best possible model of the upper atmosphere would be solely based on reliable and accurate experimental results, and would involve the application of sufficiently well established physical laws. In actual practice, due to the lack of relevant data, no purely empirical model can be established without relying on unproven (or working) hypotheses and simplifications. This even applies to the semi-empirical models of JACCHIA, which are supposed to be solely based on reliable data on satellite drag, and on undoubtedly valid law of gravitational-diffusional equilibrium, but which in actual practice had to be supplemented by other assumptions, some of which have proved incorrect (e.g., the assumption of stationary conditions at the lower boundary at 120 or 90 km altitudes, or the assumption regarding the type of $T(h)$ function etc.). Therefore it is important to understand clearly the nature of the inaccuracies involved in different models. Some of them were considered in the preceding sections.

For our purposes, the most relevant properties of atmospheric models are those involved in the analysis of the variation of the $F2$-region of the ionosphere. We shall mention once again that the following information should be available: 1) absolute values of the temperature T_∞, and of the concentrations of principal atmospheric components (N_2, O and O_2); 2) their annual variations throughout the year, with an accurate description of seasonal and semiannual changes; and 3) diurnal and latitudinal variations.

In solving various specific problems other properties of atmospheric models may also prove important. We shall now estimate the degree of accuracy of the particle concentrations and the temperature of the upper atmosphere, as described by the models.

Concentrations. The value of atmospheric models consists in their ability to describe various helio-geophysical relationships governing the variation of atmospheric parameters. A correct description of the absolute parameter values is also of importance.

An extensive review of rocket measurements /295/ revealed that the concentrations of atomic oxygen given by all models which were known prior to this publication were about one-half of their true values. This error was corrected by JACCHIA in his 1977 model.

Table 3.3 shows the data of the review /295/, the more recent rocket data /228/ for mid-latitudes, at 150 km altitude for $T_\infty \approx 900°K$, and the data contained in certain atmospheric models, tabulated in /177, 188/. The values, $F = 123$, $A_p = 4$, were assumed to correspond to T_∞-values of between 900 and 950°K. The table also includes the data of the semiannual model /24/, established for a higher temperature ($T_\infty = 1100°K$) and, for comparison purposes, the data of the daytime model CIRA–65 and the JACCHIA–77 model for similar T_∞-values. It is difficult to compare the data from different models in a simple fashion (see Table 3.3).

In the JACCHIA models published after 1970 the [N_2]-values are somewhat lower, and closer to those obtained by rocket measurements. At $T_\infty = 1100°K$ the concentration of molecular nitrogen is clearly higher than at $T_\infty = 900–950°K$. The values of the overall density ρ in the later models are also slightly lower.

The variations in the concentration of atomic oxygen [O] are more complex than variations in [N_2]. This is apparent on studying the four models of JACCHIA (65, 70, 71 and 77). The [O]-value is 1.4 times higher in the JACCHIA–71 model than in the JACCHIA–70 model. However, in the JACCHIA–77 model, as in the MSIS model, the concentrations of both atomic and molecular oxygen are lower than in JACCHIA–71. It follows that the selection of [O]-values presents certain difficulties, and the values given by any two models may differ by a factor as large as 2. It should be noted that this also holds true with respect to rocket measurements; the [O]-value in review /228/ is slightly more than one-half of the one given in review /295/.

Such discrepancies between individual [O]-values may also be found in other studies. Thus, according to Ivanov-Kholodnyi /20/ they probably occur because insufficient allowance has been made for semiannual variations in [O]. Comparison with the data from the semiannual model /24/ indicates that the more recent data in review /228/, and also the data from the 1965 models, and the data in /295/ and the JACCHIA–71 model correspond to conditions at solstice or close-to-solstice and at the equinoxes,

TABLE 3.3

Atmospheric parameters at 150–km altitude at low and medium solar activities from atmospheric models and data from rocket measurements.

Model	Reference	Atmospheric components				
		$[O]\ 10^{-10}$ cm^{-3}	$[N_2]\ 10^{-10}$ cm^{-3}	$[O_2]\ 10^{-9}$ cm^{-3}	$\Gamma 10^{-12}$ Λ/cm^3	$T_\infty K$
			Low solar activity			
Rocket data	/295/	2.30	2.60	2.50	1.96	–
	/228/	1.22	2.79	2.56	1.77	–
JACCHIA–65	/188/	1.67	3.02	3.77	2.05	900
JACCHIA–70	/188/	1.68	3.10	3.90	2.10	950
JACCHIA–71	/188/	2.36	2.54	2.68	1.95	900
JACCHIA–77	/190/	1.71	2.82	2.40	1.90	900
MSIS	/177/	1.72	2.73	2.49	1.70	941
			Medium solar activity			
CIRA–65	/175/	1.40	3.18	4.52	2.08	1150
JACCHIA–77	/190/	1.74	3.14	2.75	2.07	1100
Semiannual						
solstice	/24/	0.93	3.10	3.10	2.65	1100
equinox	/24/	2.63	3.25	3.27	3.82	1100

respectively. Recent models /177, 190/ reflect the average annual values of [O]. Thus, in selecting a suitable value of [O], semiannual and seasonal variations must be considered. In this context we shall investigate the description of annual parameters and their variations, especially for the more recent (1974–1978) models.

The following conclusions may be reached from Fig. 3.10, which represents the annual variations of [O] and [N_2].

1. The amplitude of seasonal variations of [O], i.e., the winter/summer concentration ratio, is 1.2 – 1.6. Thus, seasonal variations of [O] at 300 km altitude are smaller than at altitudes of about 100 km. The seasonal variations of [N_2] are opposite to those in [O], but are in the same direction as the changes in the temperature. In summer they have a maximum not only at 300 km altitude, but also around 120 km in all atmospheric models considered here. This finding contradicts the Groves model /169/, according to which [N_2] increases 1.3 – 1.5 times in winter in the turbopause zone, as mentioned above. Above 275 km /21/ or 260 km /176/ the sign of the seasonal variation of [O] at low solar activities is reversed.

2. The ratio between the maximum equinox concentrations and the summer concentrations of atomic oxygen obtained from the models is smaller than that given by rocket measurements. Thus, according to the rocket data (sec. 3.2.1) it is 3, i.e., $1\frac{1}{2} – 2$ times as high.

3. In all recent models represented in Fig. 3.10, the observed October peak of [O] is higher than the April peak.

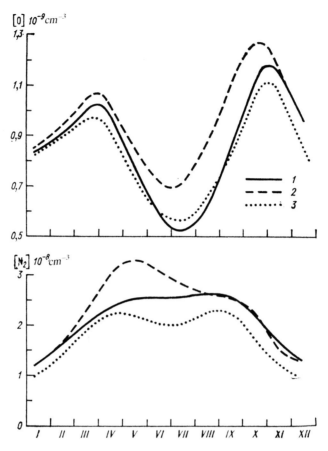

Fig. 3.10. Variations of [O] and [N$_2$] at 300 km altitude over the
year at 45° lat. under "standard" conditions ($F_{10.7} = \bar{F} = 150$; A_p
= 4; t = 14 hours) as described by three models:
1 – OGO–6 /176/; 2 – MSIS /177/; and 3 – JACCHIA–77 /190/

We must clarify the reasons for these discrepancies, which exceed the experimental
errors involved in the respective observations, and begin with the one referred to in
point 2 above.

The discrepancy between the rocket and the satellite data, concerning the relative
amplitude of semiannual variations in [O], has not yet been explained. However, we can
only discard rocket data in favor of the much more abundant satellite data if the two
types of measurements can be performed with the same degree of accuracy and
reliability. However, it is not really known which data are the more accurate, or rather
since data from both sources are known to contain major errors, which data are the less
accurate. If we choose model data as being the more reliable, it is still unclear which
model is the correct one, since even models based on identical mass-spectrometric
measurements yield estimates differing from one another by a factor of 1.4 for [O] at 300
km altitude /176, 177, 197/.

As compared with the above discrepancies, the differences between the rocket and the satellite data no longer seem as pronounced as before, but it is nevertheless of theoretical interest that the reason for the discrepancies be understood. In comparing the equinox/summer ratios obtained from rocket data with those obtained from the satellite data, in the manner described above, we failed to take into consideration the fact that this ratio generally varies with the altitude. In fact, the concentration ratio $[O]_{autumn}/[O]_{summer}$ (e.g., according to the MSIS model at 45°lat.) varies with the altitude as follows:

h_{km}	400	300	220	200	150	120
$[O]_{autumn/summer}$	1.68	1.84	2.12	2.20	2.30	2.35

The variation of this ratio with the altitude is mainly caused by the variation in the temperature T_∞ from about 1125 to about 1000°K between summer and autumn. Since the rocket data refer to altitudes of 100–130 km, and the satellite data to altitudes of 200–400 km, we find that at 120 km altitude, even according to the MSIS model, the semiannual component of variation of [O] is only lower by a factor of 1.28 than the value obtained from the rocket data; with other models, this discrepancy is even smaller.

We may note, on the other hand, that in the semiannual model /24/, which is based on rocket measurements, the ratio between equinox and summer concentrations of atomic oxygen is practically independent of the altitude between 130 and 200 km, only decreasing from 2.83 to 2.72 between these altitudes. This is explained by the fact that the same temperature, $T_\infty = 1100°K$, is assumed in the model for the equinox and for the summer. If a correction is introduced to take into account the decrease of T_∞ in autumn, as in the MSIS model, this ratio also decreases at altitudes higher than 130 km. If compensation is made for this factor, the differences between the autumn/summer ratios at the respective altitudes in the Ivanov-Kholodnyi /24/ and MSIS models would only be 20%. It is therefore justifiable to assume that there are no major discrepancies between the rocket and the satellite data, since the experimental differences are in fact 20–30% rather than 50–100%.

Many models currently used suffer from the disadvantage of being based on predictions of semiannual variations of [N$_2$] which, according to /24/, cannot be observed by rocket measurements. Different models account for this variation in different ways. Thus, in JACCHIA's model /190/ special corrections for the semiannual variations of ρ are introduced in the average annual values; however, from a purely formal point of view, all components are thus corrected to the same extent, with the result that the semiannual variations of [O], for example, are low (as shown above they are 1.5 according to /90/ and 2 according to /176, 177/), while a semiannual component appears for [N$_2$] which is an artifact. In the following section we shall attempt to clarify the sources of the error causing semiannual variations of [N$_2$] to appear in models /176, 177/, and also give a quantitative estimate of the seasonal variations in the concentration of molecular nitrogen.

In a previous section we discussed a new model /95/, based on satellite measurements of ρ, taken over several years, which are also employed in JACCHIA's model. The annual variations of [O] and [N$_2$] in the model /95/ resemble those in model /177/, so that the same comments apply to both of them.

In many aeronomic problems, and in a "non-differentiated" approach to ionosphere modeling (cf. sec. 3.4), a small semiannual component or large seasonal variations of $[N_2]$ or $[O_2]$ are often irrelevant. Thus, for instance, when calculating h_m or $n_e{}^m$, these are compensated by the respective low values of semiannual or seasonal variations of $[O]$. In isolated cases, however, it is possible to conduct a selective study on the effect of the variations in atmospheric composition on computations of ionospheric parameters; e.g., in the analysis of relative ion concentrations, the limited nature of such an approach is evident /1/.

Description of the variations in the concentration of molecular oxygen is the main difficulty in constructing models of upper atmosphere. It is apparent from Table 3.3 that the total $[O_2]$ at 150 km altitude is approximately halved in models published after 1970, thus bringing it into agreement with the rocket data /228, 295/.

When proceeding from absolute concentrations to the annual variations of $[O_2]$, it should be borne in mind /20/ that the concentration of molecular oxygen is 2–4 times higher in summer than in winter. Such seasonal variations of $[O_2]$ have not been considered in any of the models proposed for the upper atmosphere so far. This is due to the fact that according to current practice the concentration of molecular oxygen is not accurately computed in these models with allowance for dissociation and diffusion processes, as in /25/, but is merely given as a specific fraction of the total number of molecules in the atmosphere. In reality, however, this fraction is not constant, but varies during the course of the year.

Temperature. The problem of accurate determination of the temperature of the upper atmosphere was under active investigation a few years ago. It was noted that there were large differences between the T_∞-values obtained by different methods. Thus, for instance, for a given $F_{10.7}$, the daytime T_∞-values obtained by CIRA–65 model were higher by 160–200°K than those obtained by using JACCHIA–65 model. The T_∞-values obtained by the incoherent scatter method were lower than those obtained using the JACCHIA–65 model /77/. For this reason, the T_∞-value in the JACCHIA–71 model is 100–200°K lower than in the other method. Determinations of the Doppler width of the airglow line, $\lambda = 630$ nm, performed by the satellite OGO–6, indicated that the incoherent scatter data are higher by 100°K /101/.

It is apparent that the T_∞-values have undergone a considerable "devaluation" in the course of the last 10–15 years, especially for high $F_{10.7}$-values. According to Thuillier et al. /285/, the difference between the average annual values of T_∞, obtained by the optical method and the incoherent scatter method, is on the average about 1°K. The largest differences (up to 20°K) are observed in summer. More precise T_∞-values are not required in the $F2$-region calculation.

The development of the concepts concerning the magnitude of T_∞ is instructive. Satellite drag data in the 1965 atmosphere models allowed planetary distribution charts of T_∞ to be constructed, and these gave the impression of being reliable (especially to non-experts). This is why results obtained by other methods, particularly by the incoherent scatter method, were originally thought to be unreliable if they failed to agree with those given by atmospheric models. For example, the daylight temperature maximum was found by the incoherent scatter method to occur at 16 hrs /18/, rather than at 14 hours, as indicated by the models. While stationary conditions are assumed to prevail at altitudes of 100–120 km in these models, the incoherent scatter method allows for the diurnal variations of T_n in the different seasons, and for the semidiurnal variations of T_n with an amplitude of up to 50–100°K /20/. The preference for the most reliable models was based on the supposedly obvious superiority of the satellite method, which was both global and continuous, and yielded many thousands of values, as compared to the discontinuous operation of incoherent scatter stations on the ground. The difference in the accuracy of the measurements made by the two methods was ignored. It was only after mass-spectrometric measurements taken from OGO–6, ESRO–4 and other satellites had proved that the atmospheric models employed at that time gave an incorrect global distribution of T_∞, with the maximum at the subsolar point and not near the poles (an observation which was first made as a result of the determination of the Doppler width of the airglow line at $\lambda = 630$ nm), that doubts arose as to the reliability of the models employed and their inherent drawbacks became evident. It was established that this was due to the fundamental error involved in determining T_∞ from the data on the altitude distribution of ρ.

It was shown by Mayr et al. /213/ that the "temperatures" obtained from the altitude distribution of the overall atmospheric density ρ, and of the concentrations of individual atmospheric components, in particular [O], assuming gravitational-diffusional equilibrium, are not identitical with the kinetic temperature of the atmosphere. Analysis of the results of the observations performed from the satellite AEROS–A showed /112/ that the seasonal and latitudinal variations of T_∞ (O) are substantially different from those of T_∞ (N$_2$) and T_∞ (Ar), which are closer to the variations of kinetic temperature. Since at the flight altitudes of the satellites the atomic oxygen is the principal atmospheric component, the variations of T_∞ in models probably closely represent the variation of T_∞ (O). In actual fact, however, they have been computed from the experimental distribution of ρ over a range of altitudes, and for this reason differ even more from the kinetic temperature than does T_∞ (O).

It would be incorrect to suppose that, since the absolute T_∞-value is not directly involved in the calculations of the $F2$-region, it is altogether irrelevant. The value of T_∞ is an input parameter for atmospheric models, and thus variations in T_∞ directly affect the concentrations of the atmospheric components employed in calculations. As has already been pointed out, the first thing to do is to check the correctness of the annual variations of T_∞ as described by the model.

According to /95, 177, 285/, three different models yield practically identical descriptions of the asymmetric evolution of T_∞ in the course of the year, which may be described as the sum of two components (annual or seasonal, and semiannual), slightly displaced with respect to each other (Fig. 3.11). In the case of the model /285/ this may be expressed by the formula

$$\bar{T}_\infty = 1041 + 96.7 \cos\Omega(d - 168) + 23.2 \cos2\Omega(d - 98) \qquad (3.7)$$

where d is the serial number of the date, and $\Omega = 2\pi/365$ is the angular frequency of rotation of the Earth. It follows from equation (3.7) that the amplitude of the semiannual component as 24% of the annual component. In model /95/ it is 23%. Figure 3.11 shows the variations of T_∞ obtained from the models /176, 177, 192/, and the experimental curve is plotted. As may be observed from the figure, the experimental curve is more symmetrical and, for mid-latitudes, it can be described by the formula

$$\bar{T}_{\infty} = 1038 + 85\,\cos\Omega(d - 181) + 9\,\cos2\Omega(d - 106) \tag{3.8}$$

the amplitude of the semiannual component representing a much smaller proportion – only 10.5%. In the JACCHIA–71 and JACCHIA–77 models /188, 190/ only ρ has a semiannual component, this component being altogether absent in T_{∞}. Thus, in recent atmospheric models /95, 177, 285/ the semiannual component of T_{∞}-variations is much too high. This is why the variations in the concentrations of N_2, Ar and O_2 have a semiannual component which is an artifact (cf. sec. 3.2.3).

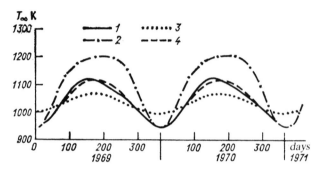

Fig. 3.11. Variations of T_{∞} during the year according to three models /95, 176, 177/ at 45°lat, under the following conditions:
$F_{10.7} = \bar{F}_{10.7} = 150$ and $K_p = 2$.
1 – MSIS; 2 – OGO–6; 3 – JACCHIA–71; 4 – experimental data obtained by the incoherent scatter method at the Millstone Hill station /256/.

Seasonal variations. According to Fig. 3.11, the model value of T_{∞} in winter is 930°K, i.e., about 200°K less than in summer. According to incoherent scatter data /256/, this difference is somewhat less (about 170°K). This large amplitude of seasonal component variations of T_{∞} brings about large seasonal variations of component concentrations. Thus, according to /188, 190/, when T_{∞} increases from 900 to 1100°C, the concentrations of N_2 and O_2, which would be constant at an altitude of about 100 km, change by a factor of 1.4 – 1.5 at an altitude of 150–170 km. Thus, whereas at altitudes of about 100 km the concentration of molecular nitrogen in winter is 1.3 – 1.5 times higher than in summer, at altitudes of 170 km and above the amplitude of seasonal variations has the opposite sign, i.e., $[N_2]$ is lower in winter than in summer. This explains the seasonal variations of $[N_2]$ in the models at 300 km altitude (point 2, (Rp) 91). Thus, any discussion of semiannual and seasonal components should take the into consideration variations of the amplitudes with the altitude, caused by the annual variation of T_{∞}. Another consequence of this situation is that inaccurate determinations of T_{∞} in models may result in distorted concentration values.

Latitudinal variations. The ratio between the amplitudes of semiannual and annual variations of atmospheric density and composition may vary with latitude. It would appear that the amplitudes of the semiannual components of ρ and [O] are independent of latitude /176, 177/, being largest at the equator and gradually decreasing towards higher latitudes /95/. The amplitude of the annual component, on the contrary, increases on passing from low to high latitudes. This is in fact observed for ρ, since the concentration of molecular nitrogen makes a major contirbution to the density, and this varies chiefly according to the annual component. All these features are clearly shown in Figs. 3.8 and 3.9. In addition to this general structure of latitudinal variations, there are also special features, characteristic of the equatorial region in addition to the asymmetry of the parameter values for the southern and northern hemispheres.

Unlike the models OGO–6 and JACCHIA–71, the MSIS model displays an equatorial bulge at equinoxes and solstices /176/, such as those mentioned in the preceding section in the context of model /95/ and Figs. 3.8 and 3.9. Marked maxima of the atomic oxygen concentrations in the sub-equatorial region were noted /124/ from the data on the emission of 557.7 nm radiation.

The amplitude of seasonal variations of the main atmospheric parameters T_∞, ρ, O, N_2 and He at $h = 400$ km is larger in the southern than in the northern hemisphere. This may be due to the fact that in the southern hemisphere the seasonal variations are enhanced by the variations of the solar flux caused by the elliptic shape of the Earth's orbit. A similar effect was noted for the critical frequencies of the E-layer /35/.

Accuracy. Until accuracy of the model parameters can be quantitatively estimated, the practical value of a model and its potential applications remains unclear. Even the authors of any given model may not be able to give an objective estimate of its accuracy; for several models no such estimates are available at all. In some cases accuracy can be estimated by comparison with other data.

In the model OGO–6 /176/ the accuracy of determination of the temperature T_∞, which is calculated from the altitude distribution of $[N_2]$, was estimated at $\pm 50°K$ on the average. This refers to the accuracy of relative determinations. With regard to the accuracy of the determination of absolute T-values, it will be recalled that Hedin et al. /176/ obtained a temperature higher than $T(N_2)$ by about 50°K, basing this on the distribution of atomic oxygen concentrations, while, in contrast, the temperature determined on board OGO–6 from the determination of the Doppler width of the 630 nm atmospheric emission line was 100°K lower /101/. The latter result is also confirmed by the incoherent scatter results. Hedin et al. /177/ used the MSIS model, in which the absolute T_∞-value had been adjusted to achieve consistency and observations made by incoherent scatter method, and showed that during the summer period the uncertainty of individual estimates of T_∞ from the model is up to 20–40°K as compared to the determinations made at Millstone Hill and St. Santin stations /256/, while the average mean square deviation was 36–42°K /177, 285/. Thuillier et al. /285/ made a comparison of T_∞-values determined by different methods, which gives some idea of the accuracy of the absolute T_∞-values. JACCHIA–71 and JACCHIA–77 models /188, 190/ do not include any estimates of the accuracy of individual determinations of T_∞.

It is apparent that the typical error of relative T_∞-determinations in models is $\pm 50°K$, i.e., about $\pm 5\%$ of the value of T_∞.

The mean square error of the determination of the overall density of the atmosphere, ρ, is estimated at about 10% by the model /95/. Clearly, this estimate also applies to the models of JACCHIA, from which the principle of constructing atmospheric models on the basis of satellite drag data was first taken. The percentage error in relative determinations of ρ is about twice as large as in determinations of T_∞, but is still quite low as compared to many other aeronomic parameters. The situation is less satisfactory with respect to the concentrations.

In the MSIS model, the errors quoted in the estimates of absolute concentrations of N_2, O and O_2, at altitudes above 150 km, are $\pm15\%$, $\pm20\%$ and $\pm50\%$, respectively /177/. It may be assumed that the errors involved in individual relative determinations of these magnitudes are of the same order, and certainly no greater. According to /177/, at altitudes above 190 km, the average square deviation of the model values from those measured in different satellites is between 14 and 23%. At the same time differences in the concentration values used in the OGO–6 model /176/ may be as high as 100%. The improvement of model /177/ as compared to model /176/ is probably due to better temperature determinations. In other models accuracy estimates of concentration determinations are altogether absent.

We may finally note that the comments made above about the drawbacks of the different atmospheric models are accompanied, wherever possible, by quantitative data and accuracy estimates. It is therefore possible to decide, if necessary, whether or not they should be taken into consideration in accurate computations and in special studies. In many cases the existing atmospheric models fully satisfy practical requirements, if these are not too exacting, and yield an approximate, but adequate description of the numerous variations in the upper atmosphere.

3.3 RATE CONSTANTS OF ION-MOLECULE REACTIONS

Of the very large number of ion-molecule reactions which take place throughout the ionosphere (see, for example, /11, 50, 223/), two reactions are of principal importance in the $F2$-region:

$$O^+ + N_2 \rightarrow NO^+ + N; \; O^+ + O_2 \rightarrow O_2^+ + O. \qquad (3.9)$$

The rate constant γ_2 is about one order of magnitude higher for the second than for the first reaction, but in the ionospheric $F2$-region the concentration of O_2 more than one order lower than that of N_2. As a result, the first reaction is more important than the second one.

In the past, the determination of the rate constants, especially that of the first reaction, proved to be very difficult, on account of its very low value as compared with those of other reactions. Many different attempts to solve this problem have been made during the past 15–20 years, but until very recently there was no general consensus on any particular solution. The main issue was the choice between direct laboratory determinations and indirect ionospheric estimates. Both methods have their undoubted advantages, but both also suffer from drawbacks. It is certainly desirable to determine the value of the rate constant under strictly controlled, suitable laboratory conditions (if this proves to be at all possible), but it is also satisfactory to obtain the value of the rate constant directly, under conditions that prevail in the upper atmosphere, on the basis of ionospheric data, making accurate allowance for all the factors which may possibly interfere (a difficult task indeed in practical terms).

Table 3.4 contains the results quoted in a number of early reviews. It is apparent that up to 1963–1965 the rate constants γ_1 and γ_2, obtained in the laboratory, differed by $1\frac{1}{2}$ – 4 orders of magnitude. Similar differences were also obtained prior to 1963 for the value of γ_1 estimated from ionospheric observations /223/. However, as early as 1965, ionospheric estimates of the constants γ_1 and γ_2 /11, 187/ from the results of rocket measurements, became both more specific and closer to the true values than those obtained in the laboratory. Even before, in 1962, Norton et al. /226/ used ionospheric data to obtain the value $\gamma_1 = 10^{-12}$ cm^3/sec, subsequently confirmed in 1973 for $T \approx$ 1000°K by the most modern and most reliable laboratory measurements to date /156, 194, 195, 215/. Table 3.4 does not include numerous other indirect ionospheric estimates of the constants γ_1 and γ_2 (those mentioned in reviews /11, 156, 158, 187, 226/ and some more recent determinations), since they are much less accurate than modern laboratory determinations.

TABLE 3.4

Values of the constants, γ_1 and γ_2 (cm^3/sec), obtained in different determinations.

Year	References	Data	γ_1	γ_2
Prior to 1963	/2.23/	Laboratory	10^{-8}–10^{-12}	–
		Ionospheric	10^{-9}–10^{-13}	–
1965	/11, 187/	Laboratory	$(5$–$200)\cdot10^{-12}$	$(0.2$–$8)\cdot10^{-11}$
		Ionospheric	$(0.5$–$3)\cdot10^{-12}$	$(0.5$–$5)\cdot10^{-11}$
1969 (1970)	/158/	Laboratory $T=300K$	$(2\pm1)\cdot10^{-12}$	$(2\pm1)\cdot10^{-11}$
1973	/156, 194, 195, 215/	Laboratory $T=500$–$1500K$	$(0.5$–$0.8)\cdot10^{-12}$	$(0.8$–$1.5)\cdot10^{-11}$

It is apparent that successful rocket studies of short-wave solar radiation and physical conditions in the ionosphere, including ion composition, temperature and other parameters, made it possible for the ionosphere experts to estimate γ_1 and γ_2 to within a factor of 2–3 much sooner than this could have been done in the laboratory. However, the technique of laboratory determinations has much improved during the past few years, owing to the development of methods such as flowing afterglow, crossed ion beams and beams of neutral molecules, and the flow drift tube. As a result, the accuracy of the laboratory measurements, even of relatively slow ion-molecule reactions, has improved greatly since 1970, and it is now possible to study such fine effects as the temperature dependence of γ_1 and γ_2, the composition of the buffer gas etc.

The problem now under intensive investigation is how to enhance the accuracy of the determinations with the aim of detecting the large number of variations in the γ-values. It was found that different workers using different measurement techniques often obtain absolute γ-values differing by a factor of 2–3, while if the same method is employed, even fine variations of the relative values are reliably detected.

Recent studies of the constant γ_1 are aimed at establishing the nature of its dependence on energetic characteristics. This research is now proceeding in two directions: (1) the effect of excitation temperature; and (2) the effect of the relative motion of the reacting species, as determined by the respective energy or by the effective temperature. We shall discuss the results from these two types of studies.

Analysis of the variation of the reaction constant

$$O^+ + N_2^* \rightarrow NO^+ + N, \tag{3.10}$$

in which the N_2 molecule is vibrationally excited, proved to be relatively simple. It was studied by Schmeltekopf et al. /259, 260/.

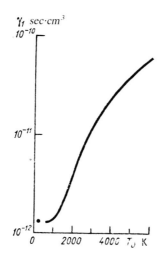

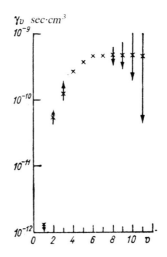

Fig. 3.12. Reaction constant γ_1, as a function of the temperature T_v, of vibrational excitation of N_2, obtained in laboratory experiments /259/.

Fig. 3.13. Constants of reaction (3.10) for excited N_2^*-molecules as a function of the excitation level, v /259/.

A well-organized accurate laboratory experiment /259/ revealed that γ_1 rapidly increases with increasing temperature of vibrational excitation of the N_2 molecule, T_v (Fig. 3.12). When T_v was raised from 1000 to 3000°K, the value of γ_1 increased by a factor of about 10; this situation may be approximately described by the equation

$$\gamma_1 = A \cdot 10^{(T_v-1000)/2000}. \tag{3.11}$$

However, as may be seen from Fig. 3.12, the increase in γ_1 slows down at high T_v-values.

The dependence of γ_1 on T_v may be transformed into dependence on the excitation level v (Fig. 3.13). It may be seen from the figure that, at the first excitation level, $v = 2$, the value of γ_v is higher by about 1½ orders of magnitude than at the ground level, $v = 1$. Thus, if the upper atmosphere contains as little as 3% of the excited nitrogen molecules N_2^*, the value of γ_1 will increase to about $1.5 \cdot 10^{-12}$ cm^3/sec, since $\gamma_{1eff} = \Sigma_v \gamma_v[N_2(v)]/[N_2(v=1)]$.

During the past few years intensive studies have been conducted on the dependence of γ_1 on the kinetic temperature and on the relative energies of motion of O^+-ions and N_2-molecules in general.

It will be recalled that, during the period around 1960, when interest was centered on the establishment of the theory of ionosphere formation, differences of opinion arose as to the value of the effective recombination coefficient in the ionosphere, α'. This problem was resolved by Biondi and his co-workers, who conducted a series of brilliant laboratory experiments aimed at studying the processes of dissociative recombination /33/. Another controversial point – whether γ_1 increased or decreased with increasing temperature /33, 158/ – arose during ionospheric investigations in the early 1970's. The matter was investigated in another laboratory experiment performed by Biondi et al. /194, 195/, and their results were soon conclusively confirmed by /76, 156, 215/.

The principal finding was that the constants γ_1 and γ_2 vary both as a function of temperature and as the relative motion of ions and electrons.

Both these factors must be considered; this is done by introducing the parameter KE_{cm} – the average relative energy of motion of an ion and a molecule bound to their center of mass in a coordinate system:

$$.KE_{cm} = \frac{m_n m_i}{m_n + m_i} \frac{u^2}{2} + \frac{3m_n kT_i}{2(m_n + m_i)} + \frac{3m_i kT_n}{2(m_n + m_i)} \tag{3.12}$$

where m_i and m_n are the masses of the ion and the molecule, respectively, and u is the average drift rate of the ions with respect to the neutral gas. Under laboratory conditions the value of u depends on the applied electric field when the measurements are carried out by the flow drift tube method. Consequently, for a suitably chosen value of u the same KE_{cm}-value may be obtained at different T_i and T_n values. The same KE_{cm}, may be used even if the distribution is not Maxwellian.

It has been shown /206, 243/ that irrespective of the temperature, accelerating electric fields and the nature of the buffer gas, the γ-value is unambiguously determined by the value of KE_{cm}. Albritton et al. /76/ successfully coordinated various earlier measurements with modern laboratory data in this manner.

St. Maurice and Torr /271/ evaluated the constants, γ_1 and γ_2, with respect to the conditions prevailing in the ionosphere of the Earth. They allowed for the fact that if $T_i > T_n$, the velocity distribution of the particles may differ from the Maxwell distribution, but assumed that

$$T_i = T_n + \frac{mu^2}{3k} \tag{3.13}$$

where the drift rate $\mathbf{u}$ in the presence of electric field E is given by the usual expression

$$\mathbf{u} = [c\mathbf{E} \times \mathbf{B}]/\mathbf{B}^2. \tag{3.14}$$

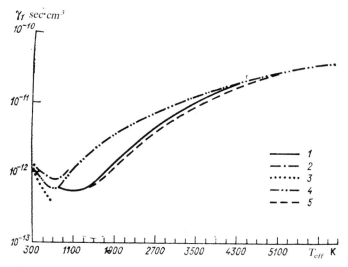

Fig. 3.14. Reaction constant, γ_1, as a function of the effective temperature T_{eff}, according to laboratory results obtained by different workers:
1 – Albritton et al. /76/; 2 – Biondi et al. /194/; 3 – Dunkin et al. 127/;
4 – McFarland et al. /215/ (at $T_{eff} = T_v$); 5 – Calculated from the data
of Albritton et al. /76/ for excitation conditions in the ionosphere /288/.

It should be borne in mind, however, that the supplementary increase in T_i in the
ionosphere is mainly caused by the interaction of the ions with the electron gas.

Figure 3.14 shows the results of the modern (curves 1 and 5) and the best known early
laboratory determinations of γ_1 /288/. It was assumed for this purpose, that depending
on the value of KE_{cm} (eV), the value of $T_{eff} = \frac{2KE_{cm}}{3} = (Ke_{cm})$ is 7740°K, and all
data were plotted on the base of this universal parameter. Curve 3 for $T_{tr} = T_v$ was
obtained by the method of flowing afterglow; curve 2 by the static drift tube method;
curve 4 was obtained by the flow drift tube method. Curve 4 represents a synopsis of the
data, which together with several others /194, 255/, were obtained by the method of
crossed ion-molecular beams, and which may be described by the relationships:

$$\gamma_1 = (1.2 \pm 0.1) \cdot 10^{-12}\left(\frac{300}{T_{eff}}\right)^{1\pm0.4} \text{ for } T_{eff} < 750K$$

$$\gamma_1 = (8.0 \pm 2.0) \cdot 10^{-14}\left(\frac{T_{eff}}{300}\right)^{2\pm0.2} \text{ for } T_{eff} > 750K$$

(3.15)

According to /291/, these relationships are identical, within the experimental error,
with the relationship $\gamma_1(T_v)$ given in /215/ (Fig. 3.12). If the experimental curves 2–4 are
compared, it is apparent, as already mentioned above, that at $T = 300°K$ the data are
practically identical, while around the minimum, at $T \approx 750°K$, they differ by a factor of
approximately two. It will be noted that at higher T_{eff}-values there are differences
between the earlier (curve 4) and the recent (curves 1 and 5) data.

Curve 5 represents the results of the calculations /288/ based on cross-sections measured by Albritton et al. /76/. The calculations take into consideration the deviations from Maxwell type distribution of ions in the midlatitude ionosphere. Curve 5 represents disturbed conditions.

It must be pointed out, first of all, that the divergence between curves 4 and 1 above $T_{eff} = 750°K$ show that γ_1 varies with T_v in an altogether different manner than it does with T_{eff}. It is important to bear this in mind, since the modern approach to the problem of variation of γ_1 with the temperature radically differs from the older concepts. Earlier observations /215, 291/ indicated that γ_1 varied with T_v and with T_{kin} in a similar manner. It was claimed, accordingly, that the excited particles play an important role in the ionosphere. However, differences between the two types of variation above 1000°K were pointed out in /76, 207, 226/, while in the ionosphere studies of Torr and St. Maurice /271, 288/ doubt was cast on the reality of any significant effect of excited N_2^* molecules on the γ_1-values in the $F2$-region.

As has already been mentioned, a number of laboratory workers reported a qualitative resemblance between the variations of γ_1 and γ_2 with the temperature (presence of a minimum, increase of γ at lower temperatures), but the absolute γ-values obtained in individual studies of this kind did not always agree. Let us consider, for example, the value of γ_1 at the minimum point, at $T \approx 750°K$. The value obtained by Biondi et al. /114, 194/ was $8 \cdot 10^{-13}$ cm^3/sec, and he explained the higher values obtained by other workers by the presence of excited particles. An even lower value ($5 \cdot 10^{-13}$ cm^3/sec) was obtained by Albritton et al. /76, 215/. The higher value obtained for γ_1 by Johnsen and Biondi /194/ may be due to an inadequate allowance for the diffusion effect. The lowest value of all ($3 \cdot 10^{-13}$ cm^3/sec) was obtained by Dunkin et al. /127/.

However, all these workers obtained the same value of about 10^{-12} cm^3/sec at $T = 300°K$. Thus, the variation of $\gamma_1(T)$, obtained by each worker was different in the temperature range between 300 and 1000°K, but they all noted that γ_1 decreased with increasing T.

This decrease of γ_1 with increasing T between 200 and 600°K was first noted in 1964–1966 /154, 158/ – an unexpected result since the rate of ordinary bimolecular reactions almost invariably increases with the temperature. Ferguson et al. /157/ offered a qualitative explanation of this effect by postulating that during a very slow ion-molecule reaction, proceeding at a sufficiently low temperature, the reacting species – the ion O^+ and the molecule N_2, for example – form a short-lived, excited, energetically unstable complex $(ON_2)^+$, undergoing decomposition into NO^+ and N, which becomes more probable the more completely the energy is distributed over the internal degrees of freedom. This probability therefore increases with decreasing temperature.

In laboratory determinations the increase of the rate constant γ with decreasing temperature is noted up to $T = 80°K$ /156, 207/. On the other hand, the conventional increase in the rate constant with the temperature is noted above 750°K and above 2000°K, for γ_1 and γ_2, respectively, since the lifetime of the complex becomes very short when the mechanism just described ceases to be operative, and the conventional bimolecular reaction mechanism takes over.

The rate coefficient, k, of a bimolecular reaction may be expressed by way of the cross section Q as follows:

$$K = \int_0^\infty Qvf(v)dv, \tag{3.16}$$

where $f(v)$ is the normalized distribution function of the particles over the velocity, v. According to Langevin's theory, in an ordinary bimolecular reaction the cross-section is inversely proportional to the average velocity, $\bar{v}$:

$$Q(v) = 2\pi e \sqrt{\frac{\alpha}{\mu}} \frac{1}{\bar{v}}, \tag{3.17}$$

where α is the polarizability of the neutral particle, and μ is the reduced mass of the complex. The magnitude Q expresses the probability of formation of the ion-molecule complex in the reaction between the ion and the molecule. As the ion approaches the molecule, an electric dipole moment is induced in the molecule. If the initial collision parameter between the ion and the molecule exceeds a certain critical value, the ion will only be slightly deflected from its path of motion and no chemical reaction will take place. Otherwise, the ion is attracted to the molecule with the induced electrical charge and a complex is formed, in which the chemical reaction – i.e., redistribution of the atoms and of the energies – takes place. Clearly, the higher the gas temperature, the higher the average relative velocity of the particles, and the lower the probability of an ion-molecule reaction. In formula (3.17) this is expressed as the inverse proportionality between Q and $\bar{v}$. If (3.17) is substituted in (3.16), k is either found to be constant or to increase proportionally with $\sqrt{T}$. Thus, contrary to the situation at low temperatures, the rate coefficient k increases with increasing temperature. In the particular case of the reaction $O^+ + N_2 \rightarrow NO^+ + N$, k increases with temperature above 750°K.

Excited N_2^* molecules in the $F2$-region. We have seen that, in addition to reaction (3.9), the reaction (3.10) may take place in the ionosphere. The rate constant of the latter reaction is higher by one or more orders. In this context it is important to estimate the concentration of vibrationally excited N_2^* molecules in the ionosphere. It is determined by the balance between the opposing processes of excitation of N_2 and quenching of N_2^* molecules. The excitation levels are determined by the vibrational level, v. A particularly effective quenching mechanism is the reaction

$$N_2^* + O \rightarrow N_2 + O, \tag{3.18}$$

with a rather high rate constant at $v = 2 - 1.07 \cdot 10^{-10} \exp(-69.9/T^{1/3})$ cm^3/sec /50/.

Walker et al. /299/ were the first to show that this reaction effectively reduces the number of excited N_2^* molecules in the ionosphere. The vibrational temperature T_v, which, in the absence of reaction(3.18), should be equal to the electron temperature T_e (if the medium is excited by electrons), decreases and tends to assume the value of the kinetic temperature of atoms and molecules. Additional computations, which confirm that T_v and T_{kin} should be roughly equal below the maximum of the ionospheric F-layer, were performed by /118, 216/. Vlasov et al. /8/ recently recomputed the time period within which the excited N_2^* molecule returns to the state of equilibrium Boltzmann distribution under conditions prevailing in the ionosphere. Up to the altitude of 250 km this occurs in less than 25 minutes, i.e., the distribution of N_2 molecules over the vibration levels is essentially of the Boltzmann type.

The number of the excited N_2^* molecules in the ionosphere under ordinary conditions may thus be calculated by using the Boltzmann formula. Let the concentration of N_2 molecules at $v = 2$ be n_1, and the concentration of N_2 molecules at $v = 1$ be n_0. Since ΔE, the energy spacing between the vibrational levels of N_2 is 0.29 eV, their relative concentrations are given by the formula:

$$n_1/n_0 = e^{-\Delta E/kT} \tag{3.19}$$

The relative concentrations of the excited N_2^* molecules as a function of the temperature T_v are shown in the following table:

T_v K	600	800	1000	1200	1400	1600	1800	2000
n_i/n_0	$4 \cdot 10^{-3}$	$1.5 \cdot 10^{-2}$	$3.5 \cdot 10^{-2}$	$6 \cdot 10^{-2}$	0.11	0.122	0.154	0.186

This means that, due to the low temperature of the atmosphere by night (also in the daytime in the winter), no significant effect of the reaction (3.10) on the value of γ_{1eff} is to be expected. Such an effect is only possible either during the day in the summer, or in the auroral zone, where the temperature of the atmosphere may exceed 1000°K. It may be noted that conditions are more favorable to the excitation of O_2 molecules, since the energy of the first vibrational level $\Delta E \approx 0.2$ eV, so that the relative concentration of the excited molecules $[O_2^*]/[O_2]$ is larger than $[N_2^*]/[N_2]$.

Clearly, if the conditions are disturbed to an exceptional extent, the concentration of N_2^* molecules may become higher than the Boltzmann concentration for a limited period of time. However, even observations made during the period of stable auroral red arcs /221, 222, 265/ showed that the value of γ_1 in the F-region increased only 7.6 times, i.e., the concentration ratio $[N_2^*]/[N_2]$ was only 0.15 – 0.2. Such an increase can be simply explained by postulating a higher kinetic temperature of the F2-region (2000°K) during the period of auroral activity.

In the context of the above considerations it should be noted that since the relaxation times of vibrationally excited levels of O_2 and N_2 are finite, it is possible, in principle, for the vibration excitation temperature T_v to be higher than the kinetic temperature T_{kin}. However, since the effective vibrational relaxation times are much longer than the effective lifetimes of vibrationally excited molecules on account of other processes /8/, the discrepancy is small and may be ignored to a first approximation.

3.4. APPROBATION OF A SYSTEM OF AERONOMIC PARAMETERS

The main input parameters used in calculating the F2-region can be placed into three different groups: 1) the total short-wave solar flux; 2) the model of the neutral atmosphere; and 3) the rate constants of the principal processes. All three types of parameters are closely interconnected and should form a single, self-consistent system. Consequently, selection of parameters is a very complex task, for which there is generally moe than one possible solution. However, the solution of the problem may be approached by comparing the results of simultaneous determinations of several ionospheric parameters with the results of the calculations. Such determinations may include, for example, incoherent scatter data, and results of f_oF2 determinations obtained at the vertical sounding stations.

In order to find the solution, the $F2$-region must be studied under sufficiently homogeneous conditions. We shall begin by considering the seasonal variations around noontime. This is convenient, since such variations always take place according to a single, unchanging mechanism, whereas a description of diurnal variations, for example, must take into consideration the difference in the mechanisms of formation of daytime and nighttime ionospheres, and gives rise to more complex computations.

It is common practice to implicate three main processes in $F2$-region formation: the photoionization of atomic oxygen, the transport of O^+-ions and their conversion as a result of ion-molecule reactions. The mathematical tool employed for this purpose is the diffusion equation in the form (2.19). The input parameter for this calculation $-T_e$, T_i and the plasma flux at the upper boundary – are experimetnally determined. It may be seen from the discussion that follows below that the vertical drift around noontime undergoes small seasonal variations throughout the year, which may be corrected for in the calculations; however, 10 m/sec may be taken as the average value. The rate constants of the ion-molecule reactions, γ_1 and γ_2, which are required to calculate the coefficient of linear recombination $\beta = \gamma_1[O_2] + \gamma_2[N_2]$, are assumed to have the values quoted in secs. 2.1.2 and 3.3.

As an example, we shall demonstrate the process of selection of a self-consistent parameter for the solar flux and model of the neutral atmosphere /31/. Under the conditions assumed the principal criterion is whether or not the thermospheric model selected is a faithful description of the seasonal and semiannual variations in the $F2$-region around midday. We shall discuss the parameters of the maximum $F2$-layer; we shall than select the atmospheric model by comparing the experimental variations of n_e^m and h_m in the course of the year with those calculated by the different thermospheric models.

Average values of log n_e^m and h_m for the four seasons of the year, obtained from five different models

Parameter	Month	JACCHIA–71	JACCHIA–73
lg n_e^m	XII, I	5.79	5.57
(lg n_e^m)–lg n_e^m	VI, VII	0.04	–0.13
(lg n_e^m)$_{winter}$–lg n_e^m	IV	–0.16	–0.34
	X	–0.18	–0.20
h_m KM	XII, I	262	245
(h_m)$_{winter}$–h_m,KM	VI, VII	–30	–50
		–33	–50

Table 3.5 shows the experimental values of log n_c^m and h_m as compared to those calculated by the different models.

The averaged values of lg n_c^m and h_m, obtained by the incoherent scatter method at Millstone Hill during a period of enhanced solar activity in 1968–1970, with $F_{10.7}$ close to 150 /145, 146, 148/ will be used as the experimental data. Since the n_c^m values for autumn and winter are practically identical with the average data obtained by ionospheric sounding /73/ for the same geographic point at the Wolf number W = 100, which is equivalent to $F_{10.7}$ = 150, these data are to be preferred. It should be noted that the data in /73/ have been selected for a certain level of solar activity and have been statistically processed, and in this sense may be said to be more reliable than the above averaged values of n_c, which are based on a small series of sporadic observations carried out during the period 1968–1970.

Calculations of the $F_{10.7}$ = 150 level of solar activity under quiet geomagnetic conditions (A_p = 4; φ = 40°N; 14 hrs) involved the use of the following models: JACCHIA–71 /188/, JACCHIA–73 /189/, ESRO–4 /197/, OGO–6 /176/ and MSIS /177/.

Table 3.5 shows the values of log n_c^m for the four seasons of the year. The tabulated data were calculated for the short-wave solar flux, assumed to be as in /179/. As mentioned in sec. 3.1, the absolute value of this flux, I_o, is low, with the result that the n_c^m-values are low as well, but the relative variations of n_c^m are independent of the selected value of I_o. The vertical drift rate w undergoes minor seasonal variations – between –5 m/sec in summer and –15 m/sec in winter – the average value of –10 m/sec being assumed at equinox. The drift corrections were introduced using relationships (2.41) and (2.42).

We shall begin our discussion of the data presented in Table 3.5 with the relative variations of n_c^m, particularly those between summer and winter. It was experimentally shown that in winter the values of log n_c^m are 0.2 – 0.35 higher than in summer. The value for the anomalous seasonal variations, is about 0.3, and was obtained using all modern models (ESRO–4, OGO–6 and MSIS), while the earlier models of JACCHIA yield incorrect seasonal variations. In April and in October, during the

TABLE 3.5

and corrected for w, as compared with the experimental values.

Model				Observed
ESRO–4	OGO–6	MSIS	according to /73/	according to /145, 146,148/
5.90	5.73	5.84	6.12±0.08	6.08±0.15
0.30	0.29	0.28	0.35	0.20
0.20	–0.03	–0.02	0.20	–
–0.02	0.13	–0.06	0	–0.05
247	280	275	–	270
–40	–32	–27	–	–30
–37	–35	–10		–32

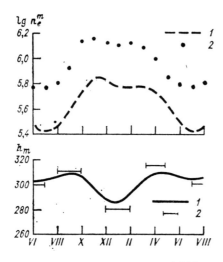

Fig. 3.15. The values of lg n_e^m, calculated for the solar radiation flux according to /179/ (1), and the experimental data (2), as well as calculated (1) and observed (2) h_m-values.

peaks in the yearly evolution of [O], the comparison does not yield clear results but, as before, the JACCHIA models are the least satisfactory. Of the latest models, ESRO–4 gives the best agreement with the experiment, MSIS satisfactorily describes autumn conditions only, while OGO–6 leads to an inverse relationship between the spring and the autumn values of lg n_e^m.

The relative variations of the altitudes h_m over the year are best described by the models OGO–6 and JACCHIA–71; the MSIS model does not give a satisfactory value at the equinox, while the ESRO–4 is not very satisfactory for the summer period. However, with regard to the parameter h_m the differences between these models are minor. Since the accuracy of the relative altitude measurements is about ±10 km, we may conclude that it is difficult to give preference to any one model on the strength of the annual variations of h_m alone. With respect to the absolute altitudes, only the OGO–6 and the MSIS values are close to those obtained by observation.

Table 3.5 lists values close to the experimental ones, given by each model. As a result of this comparison it may be concluded that, if the initial assumptions are valid, none of the models is fully satisfactory. However, the newer models (ESRO–4, OGO–6 and MSIS) show a better agreement, in one way or another, with the experimental data in the relative variations of n_e^m and the absolute value of h_m.

Let us now consider the problem of selecting the short-wave solar flux. The electron concentration in the layer maximum is proportional to the photoionization rate, i.e. to the short-wave solar flux. Thus, after the atmospheric model and the vertical drift rate are chosen, a comparison of the calculated absolute value of lg n_e^m with the experimental values will yield the total solar flux.

Figure 3.15 shows the median values of lg n_e^m for 40°lat. N., 14 hours and $F_{10.7} =$ 150, according to /73/, and the lg n_e^m values calculated according to the OGO–6 model, with allowance for the intensity of short-wave solar radiation according to /179/. The calculation was carried out on the assumption that the drift is directed downwards, and is constant at 10 m/sec throughout the year, in order to ensure the consistency of the h_m-values, which are also shown in the figure. Clearly, in order to

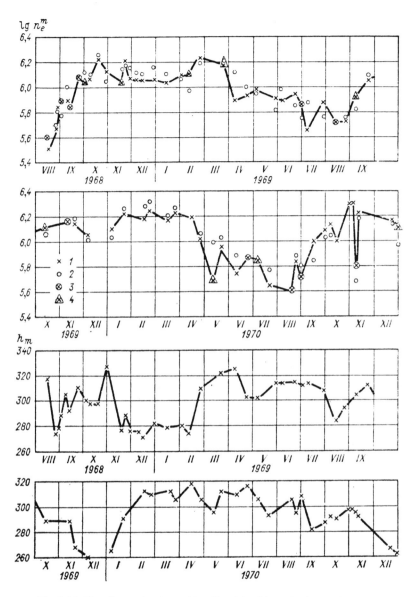

Fig. 3.16. The observed values of lg n_c^m and h_m (1), and calculated values of lg n_c^m (2) (h_m values agreeing with those actually observed were calculated by suitably selecting the rate of the vertical drift). The data (3) refer to disturbed days ($A_p \geqslant 15$), and the data (4) refer to the days immediately following such days.

obtain agreement with the experimental data for lg n_c^m, the selected value of the solar flux must be higher by a factor of about 2.2 than the one quoted in /179/. We may note that this increase by a factor of 2.2 is in agreement with the selection of molecular nitrogen, molecular oxygen and atomic oxygen concentrations appropriate to the model OGO–6, and with the vertical drift rate. It may be seen from Table 3.5 that if MSIS or ESRO–4 models are employed, the factor by which the solar flux value must be multiplied becomes smaller.

The spectrum of solar radiation has little effect on the calculation of the electron concentration in the $F2$-region, since ionizing radiation of different wavelengths reaches the maximum layer without a significant absorption, and it is sufficient for this reason to consider the variation of the total radiation flux only.

The discussion just given makes it possible to construct a closed, self-consistent system of aeronomic parameters as a basis for the calculations of the $F2$-region.

In order to verify the conformity of the system of aeronomic parameters thus selected with the real situation in the $F2$-region, as was done in /51/, calculations were conducted for 67 dates during the period 1968–1970, and were compared with the results of the observations conducted by the incoherent scatter method at the Millstone Hill station /145, 146, 148/. The observations had been taken in all seasons, and at varying levels of solar and geomagnetic activities. The calculations were based on the OGO–6 model. The spectrum of solar radiation and its variations with the activity were taken according to /25/, and a correction factor of 2.2 for the integral solar flux was introduced.

In the selection of the parameter system described above it was assumed that a downward plasma drift of $w = -10$ m/sec prevails in the $F2$-layer maximum, which is generally true under average conditions. However, regarding calculations made for individual dates it must be borne in mind that the values of w may vary from day to day. This is because the rate of the vertical drift is determined both by thermospheric winds and by electrical fields, while the diurnal variations of either of these are not known with any degree of accuracy. A procedure was accordingly adopted, which is based on the fact that the altitude of the h_m layer is highly sensitive to vertical drifts (cf. eq. (2.42)). By selecting the value of w the calculated and the experimental values of h_m are made to agree. A comparison of the electron concentration n_c^m calculated for this w-value with the concentration actually observed gives an idea of the extent to which the model corresponds to the real situation.

As an independent check, a comparison of the calculated and observed data for the overall integral plasma velocity v_z in the layer maximum may be made (the observed values of $v(h)$ are available for several dates in 1970). It is seen that the calculations may be checked against three ionospheric parameters: h_m, n_c^m and v_m, thus making a rigorous quantitative control of aeronomic input parameters possible.

Figure 3.16 shows the experimental data obtained for n_c^m and h_m over three years. It is clear that the yearly variations of the experimental values of these parameters displays, on the average, well known trends – the peak n_c^m-values occur during equinoctial and winter periods, while the minimum values of this parameter occur in summer. This so-called seasonal anomaly of the $F2$-region will be discussed in detail in Chapter 4. With respect to the altitude of the h_m-layer, an opposite trend is evident – the summer and the equinox values are higher than the winter values. However, a different picture is obtained on considering specific days. Within a single month, during magnetically quiet days, the n_c^m values may differ by a factor of more than two (e.g., 27th August and 27th September 1968), while the difference Δh_m

between these dates was 37 km. A similar situation occurred on the 5th and 26th of February 1969, when Δh_m = 30 km. Such large variations of the parameters in the F2-region are due to changes in the solar activity; in the former case $F_{10.7}$ had increased from 112 to 160, in the latter from 143 to 199.

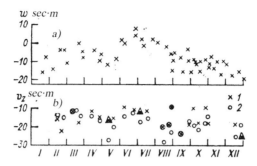

Fig. 3.17. Calculated rates of vertical drift w on quiet days (a), and calculated (1), and observed (2) velocities of the overall vertical motion of the plasma v_z in the maximum of the layer (b) on certain dates in 1970. Point markings as in Fig. 3.16.

We may inquire how effectively this calculation method reflects such variations in the input parameters. It is seen from the figure that, except for the perturbed days, the difference between the calculated and the experimental values of lg n_e^m does not exceed ±0.1. The mean square deviation of Δlg n_e^m on magnetically quiet days during the three-year period is about 8%. The scatter of ±0.1, obtained for lg n_e^m, roughly corresponds to the experimental error in the determination of this parameter. The accuracy is determined both by the experimental method employed and by the accuracy of the lg n_e^m-values read off the graphs representing the experimental material. Consequently, we may conclude that the computation system described above yields sufficiently accurate n_e^m-values on magnetically quiet days.

We shall now discuss vertical plasma velocities. We may note that the vertical drift w, determined by the system of thermospheric winds, and by the electric fields, is the third most important factor in daytime (after the effects connected with solar radiation and the composition of the neutral atmosphere) which is responsible for the variations of the altitude of and electron concentration in the layer maximum. However, as was seen in sec. 2.1.2, it is difficult to make accurate allowance for the day-to-day variation of w; and this may be done by selecting the value of w on the basis of the layer altitude, h_m, which also solves the basic problem of the agreement between the description system employed and the real situation in the daytime F2-region. On the other hand, data obtained for w are of interest in their own right since the individual, irregular, day-to-day variation of w may serve to indicate certain general relationships.

Fig. 3.17 is a synopsis of the w-values calculated on quiet days. We note a seasonal varation of w, positive values (up to 10 m/sec) and maxima during the summer months, negative values and minima (up to -10 m/sec) in winter. The average annual value of w is about -10 m/sec, as was assumed in selecting the parameters above.

We shall compare the calculated and the observed /148/ integral rates of plasma motion v_z, shown in Fig. 3.17b. The integral velocity consists of the diffusional velocity and the rate of vertical drift, and is determined in incoherent scatter installations (the accuracy of determination of v_z around 450 km in daytime is ± 4m/sec /140/.) It may be seen from Fig. 3.17b that, generally, the calculation follows the day-to-day variations of v_z. In some cases, however, the absolute differences are larger than the experimental error involved in the determination of v_z. We may also note that the calculated v_z-values tend to be somewhat low. This means that in certain cases the diffusion coefficient or the vertical drift rate do not correspond to their real values. In fact, the overall rate of motion of the plasma within the layer maximum may be expressed as $v_z = -D_m A_m + w$. It follows that the difference may be related to the value of D_m or of w, since A_m (cf. eq. (2.19)) depends on the experimentally determined values of T_e and T_i.

We may conclude from the above considerations that the system of aeronomic parameters proposed here corresponds on the whole to the real conditions in the $F2$-region around noontime. It should be clearly understood that the above approach to the selection of the parameter system was merely an example, and that a separate critical study of such a selection must be conducted for different problems, such as nighttime conditions, low altitudes etc. with allowance for the specific mechanism of formation of the $F2$-region. In the calculations of the regular variations of the $F2$-region, which will be given in the following chapters, we shall use the OGO–6 model together with the thermospheric MSIS model, since the latter model yields an effective description of noontime conditions in the $F2$-region at elevated levels of solar activity.

The MSIS model has a number of advantages (sec. 3.2), relevant to the material to be considered in the text that follows. Since this model is based on much more abundant experimental material, obtained at different latitudes and at different solar activity levels, the dependence of the composition and temperature of the upper atmosphere on these parameters is better represented by the MSIS than by the OGO–6 model. Moreover, the use of a single model to describe various helio-geophysical conditions is very convenient in practical work.

CHAPTER 4. CALCULATION OF THE REGULAR VARIATIONS OF THE
MID-LATITUDE *F*2-REGION

In this chapter we shall deal with the regular variations of the parameters of the *F*2-layer of the ionosphere, which are typical in magnetically quiet conditions. These include: the annual variations of frequencies, f_oF2, at noon; and the seasonal and the December anomalies, the dependence of the annual variation of f_oF2 on the latitude and on the solar activity level. All these problems are primarily connected with the annual variations in the *F*2-region. Diurnal variations are also regular. We shall discuss the role played by the individual processes in the formation of the *F*2-region in daytime and at night. In particular, we shall concentrate on the problem of formation of the nighttime *F*2-layer and shall discuss the phenomenon of the nightly increase of electron concentration, which is observed in winter.

The main objective of this chapter is to illustrate the possible methods and the procedures that may be developed to calculate the *F*2-region, and to interpret the observed regular variations in the layer maximum in terms of these. This review, together with the results of comparison with the experimental data, should indicate to what extent the proposed calculation method in fact describes the *F*2-region under specific helio-geophysical conditions, and whether or not it can be employed for prediction purposes.

4.1 ANNUAL AND LATITUDINAL VARIATIONS

Initial observations showed that the *F*2-region cannot be described by the theory of Chapman's single layer, and that a number of its specific behavioral features could be described as anomalies /70/. Annual, diurnal, and latitudinal variations are all anomalous including the well-known "seasonal", "December", "equatorial" and "diurnal" anomalies.

Figure 4.1 shows the annual variations of midday and midnight median values of the critical frequencies of the *F*2-layer during the various phases of solar activity cycle, obtained at the Washington station. It is seen that the around-midday values of f_oF2 vary in a complex manner during the year, and reach their peak values around equinox and their minimum values in summer. As a rule, f_oF2 is smaller in summer than in winter. The difference between the winter and the summer values of f_oF2 is particularly marked in the years corresponding to peak solar activities, and decreases as the minimum phase of solar activity is approached. The fact that the winter values of critical frequencies are higher than the summer values, which cannot be explained in terms of the simple Chapman layer, is known as the "seasonal anomaly" of the F2-layer. The effect is anomalous, since according to the theory of the simple layer the electron concentration in the layer maximum usually increases as the zenith angle of the Sun decreases, whereas according to observation n_c^m is smaller is summer and in winter. It may be seen in Fig. 4.1 that the effect of seasonal anomaly does not apply to the midnight f_oF2 values, and that the summer concentrations of electrons in the layer maximum are higher than the winter values at all stages of the solar cycle. This property of the nighttime F2-region is not related to the seasonal anomaly of the noon f_oF2-values, but rather to the generation of the nighttime F2-layer (cf. sec. 4.2).

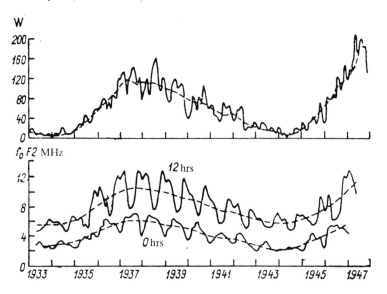

Fig. 4.1. Variation of noontime and midnight median values of f_oF2 at different phases of the solar activity cycle W, as reported by the Washington station.

A second anomalous feature of the annual variations of f_oF2 is the presence of semiannual peaks, which may be seen in Fig. 4.1 to be more or less distinct in individual years. An analysis of the global distribution charts of the critical frequencies of the F2-layer shows that these peaks may be clearly seen at one station and be altogether absent at another, depending on the latitude of the station and the level of solar activity.

The variation of f_oF2 and the parameters of the F2-layer during the year although variegated in nature, with respect to the latitude and the solar activity cycle now seem to be well understood.

4.1.1 Seasonal Anomaly. This concept usually means that the winter daytime critical frequencies are larger than the summer ones. The value of this difference varies for the different stations and depends on the geographical coordinates of the observation point and on the level of solar activity.

Figure 4.2 shows a chart of planetary distribution of the seasonal anomaly, plotted from the data furnished by 140 stations in the year 1958 during which the level of solar activity was exceptionally high (W = 200) /289/. It is apparent from the figure that the maximum seasonal differences occur at mid-latitudes in the Northern hemisphere. They are much less distinct in the Southern hemisphere, and summer f_oF2 values are higher than the winter ones, i.e., the seasonal anomaly is absent over extensive areas. A similar situation is noted at low latitudes and in the equatorial zone, where this anomaly is either very weak or altogether absent. As the level of the solar activity decreases, the areas in which the seasonal anomaly may be observed decrease in number; finally, during the years of the minimum solar activity only small areas remain in the Northern hemisphere, where the winter f_oF2 values are still larger than the summer ones (e.g., at the Washington station) (Fig. 4.1). As a rule, the seasonal anomaly is not observed in the Southern hemisphere, and the annual variation is merely expressed by the presence of semiannual maxima at the equinoxes.

It may be seen from /289/, as well as from numerous other observations performed at ionospheric sounding stations, that the annual variations of the f_oF2-values around midday display the following characteristic features. The *F*2-region is distinguished by a seasonal anomaly, which is more conspicuously manifested a) at middle and middle-to-high latitudes; b) at high solar activity; and c) in the Northern hemisphere. During equinoctial periods semiannual peaks of f_oF2 appear, which are more marked a) at low and equatorial latitudes; b) at the minimum of the solar activity, and c) in the Southern hemisphere.

The seasonal anomaly of the *F*2-region may be interpreted by assuming several different mechanisms. Thus, based on the fact that plasma diffuses along the lines of magnetic force, we may assume a plasma exchange between the magnetically conjugated points, when the plasma flows from the summer (i.e., hotter) hemisphere to the winter (i.e., colder) hemisphere, thus enhancing the winter concentration of the electrons in the *F*2-layer /5, 252/. Recently, exchange fluxes between the *F*2-region and the overlying protonosphere were determined by the method of incoherent back-scatter. Data obtained at Millstone Hill station indicate that the daytime flux, directed upwards, is $5 \cdot 10^7$ cm^{-2}·sec^{-1} on the average /147/. At a similar station at Arecibo the daytime flux values are somewhat higher – between $1 \cdot 10^8$ and $4 \cdot 10^8$ cm^{-2}·sec^{-1}. It was also found that the overall diurnal wintertime flux was directed downwards (into the ionosphere), while being generally directed upwards in summer /294/. This result seemingly confirms the hypothesis of plasma transport between the hemispheres, but in actual fact the flux values are very small; moreover, in daytime the flux is directed from the *F*2-region to the protonosphere, both in summer and in winter, and thus cannot constitute a quantitative explanation of the seasonal anomaly.

According to another hypothesis /13, 37, 113, 186, 199, 208, 214, 253/, the composition of the upper atmosphere is subject to seasonal variations, with consequent variation in the ratio between the atomic and the molecular components at the altitudes of the *F*2-region. This change in composition in turn results in changes in the rates of formation and recombination of electrons which, as shown in /249/, may explain the fact that winter n_e^m concentrations are higher than the summer concentrations. This hypothesis was subsequently confirmed experimentally by studies on

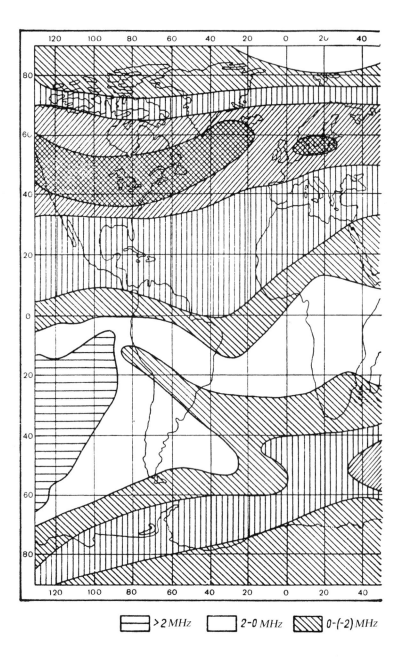

Fig. 4.2. Planetary distribution of the seasonal anomaly at the time of peak solar activity
in 1958 (differences between the summer and the winter critical frequencies in MHz)
/289/.

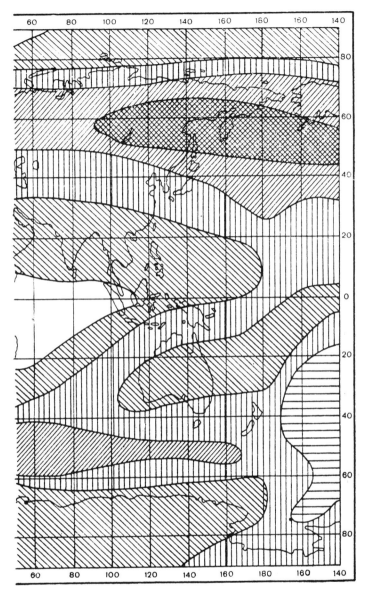

[||||] -2-(-4)MHz [////] -4-(-6)MHz [▨▨] <-6 MHz

variations of the neutral composition, using rockets, satellites and incoherent scatter installations /77, 111, 176, 257/.

A number of other mechanisms were also proposed. These, together with the variations in the composition, may provide a more complete quantitative explanation of the seasonal anomaly. Thus, Strobel and MacElroy /274/ considered the seasonal variations of the vibrational temperature of N_2 molecules, which determines the rate constant γ of the ion-molecule reaction $O^+ + N_2 = NO^+ + N$ – the principal reaction taking place in the $F2$-region. However, calculation shows that if the assumptions regarding the seasonal variations of γ are valid, this effect should be insignificant.

Ivanov-Kholodnyi and Mikhailov /29/ made a similar attempt to take into account the variations both in the composition and in the γ-value at the same time. They used the variation of the vibrational excitation of N_2, i.e., of the γ-value, with the electron temperature. The results of the calculations agreed with experimentally determined seasonal variations of n_e^m, but the evolution of the layer maximum was contrary to the observations, since the relationship $\gamma(T_e)$ gave high values for the coefficient of linear recombination.

Kohl and King /202/ attempted to explain the effect of seasonal anomaly as due to thermospheric winds, but other workers /128, 244, 274, 278, 292/ showed that the downward drift produced by these winds is stronger in winter than in summer, and thus reduces the concentration of n_e^m during the winter, so that the thermospheric winds would be expected to have just the opposite effect. On the other hand, these winds strongly affect the altitude of the layer maximum, and probably should be taken into account when explaining the seasonal variations of h_m /130, 131, 244, 274/.

During recent years improved thermospheric models (ESRO–4, OGO–6, MSIS) have been developed, more accurate values have been obtained for the rate constants of ion-molecule reactions, the values of UV solar fluxes have been reviewed, and complete data on the $F2$-region have been obtained by the incoherent scatter method. All this makes it possible to compare rigorously the experimental with the calculated data for the various parameters of the $F2$-region, and to re-evaluate the role played by the individual mechanisms in the generation of the observed seasonal variations of the $F2$-region.

Before considering the planetary characteristics of the annual variations of f_oF2, we shall study the seasonal anomaly of the mid-latitude $F2$-region, and analyze the role played by individual mechanisms in the seasonal variations of both f_oF2 and h_m. We shall consider the midday conditions prevailing at higher levels of solar activity ($F_{10.7} = 150$) at a point with the coordinates, $\varphi = 40°N$ and $\lambda = 75°W$. Table 3.7 shows the experimental data obtained at this site, which show that n_e^m is higher in wintr than in summer, and that the opposite is true for h_m. We shall take a pair of specific data, determined by observation at Millstone Hill in winter and in summer, respectively, /146/, and use them in the calculation of the $F2$-region by the method explained above.

Figure 4.3 shows the profiles obtained on typical dates, viz.: 16th January 1969: $F_{10.7} = {}^{158}/_{159}$, $A_P = 10$ and 5th June 1969: $F_{10.7} = {}^{173}/_{154}$, $A_P = 6$. It is seen that the values of h_m and $\lg n_e^m$ on these dates are close to the average values given above, and may thus be considered as being typical under these conditions.

A full explanation of the seasonal anomaly in the *F*2-layer maximum must make due allowance for three factors: the electron concentration n_e^m, the altitude of the maximum, h_m, and the depth of the layer. It may be seen in Fig. 4.3 that the winter value of n_e^m is about 1.8 times higher than the summer one, while the altitude of the layer maximum in summer is about 25 km higher than in winter. Moreover, the summer profile is wider, so that the winter concentrations are larger than the summer concentrations in a limited altitude range only – in the vicinity of the layer maximum. Figure 4.3 shows the $n_e(h)$ profiles theoretically calculated for these helio-geophysical conditions, their qualitative formation is in agreement with the experimental profile.

After the information required has been obtained by calculation, we are in a position to appreciate the role played by individual aeronomic parameters in the seasonal differences. We therefore employ the approximate relationships (2.45) and (2.46) to calculate the electron concentration and the altitude, h_m, of the layer maximum in the form of finite differences, as was done in para. 2.3.1:

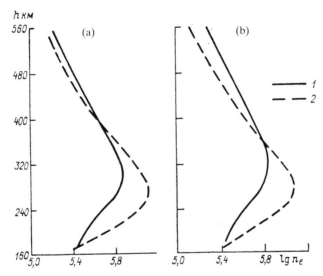

Fig. 4.3. Observed /146/ (a) and calculated (b) distribution of
electron concentrations in daytime *F*2-region on the 5th of June
1969 (1) and 16th January 1969 (2).

$$\Delta \lg n_e^m = 1.08\Delta \lg [\mathrm{O}] - 0.65\Delta \lg \beta + \lg I_1/I_2 + 9 \cdot 10^{-3} \Delta w \qquad (4.1)$$

$$\Delta h_m = 50\Delta \lg [\mathrm{O}] + 50\Delta \lg \beta + 1.55\Delta w \qquad (4.2)$$

In subsequent analysis, data on the composition, temperatures and rates of vertical drift during the two seasons are required; these are listed in Table 4.1. The table also gives the values of the coefficient of linear recombination $\beta = \gamma_1[\mathrm{O}_2] + \gamma_2[\mathrm{N}_2]$, and the meridional and zonal components of the wind, v_{nx} and v_{ny}, respectively. All parameters refer to the altitude of 300 km.

TABLE 4.1

Values of the aeronomic parameters for the calculation of seasonal variations of n_e^m and h_m.

Season	$[O]\cdot10^{-8}$	$[N_2]\cdot10^{-8}$	$[O_2]\cdot10^{-6}$	T_n	T_i	$\gamma_2\cdot10^{12}$	$\gamma_1\cdot10^{12}$	$\beta\cdot10^4$	v_{nx}	v_{ny}	w
		cm^{-3}			K	cm^{-3}/sec		sec^{-1}		m/sec	
Winter	9.11	1.30	3.83	1045	1070	1.02	7.76	1.62	−21.70	−36.00	−8.89
Summer	7.52	2.90	10.00	1185	1213	1.31	7.10	4.51	−2.10	−28.00	−2.73

The vertical drift w of the plasma induced by thermospheric winds is calculated with allowance for magnetic dip I and inclination D, 72° and −15° respectively, for the Millstone Hill station. It may be seen from the table that the seasonal variations in the concentration of atomic oxygen at 300 km altitude, calculated by the MSIS model, are small – only about 20% – while the concentration of the molecular components varies by more than a factor of two in the opposite direction. Moreover, as the summer temperature is higher than the winter temperature, the rate constant of the ion-molecule reaction, $O^+ + N_2 = NO^+ + N$ increases by a factor of 1.3, as a result of which the coefficient of linear recombination, β is 2.8 times larger in summer. Calculations indicate that the velocities of thermospheric winds in winter are higher than in summer, which means that the vertical drift w is faster in winter than in summer.

Table 4.2 shows the calculated contributions of seasonal variations of individual aeronomic parameters: the atomic oxygen concentration [O], the loss coefficient β, and the vertical drift w, the seasonal variations Δ(winter–summer) of the altitude h_m, and electron concentration n_e^m in the $F2$-layer maximum. The term lg I_1/I_2 is a correction for the different solar activity levels on the days when the observations were made.

It is seen from Table 4.2 that the seasonal variations of the $F2$-layer are mainly due to the seasonal variations of the linear recombintion coefficient β, which determines seasonal changes of n_e^m and h_m in opposite directions. The variations of [O], with the winter concentrations being larger than the summer concentrations, intensify the seasonal differences in electron concentration n_e^m and attenuate the variations in the layer altitude, h_m. The contribution of the vertical drift to the variation of n_e^m is small, but the seasonal changes of h_m are strongly affected by this factor, because the effect of the vertical drift on h_m and n_e^m is opposite to that of atomic oxygen.

Let us finally consider the third feature of the seasonal anomaly, viz., the limited altitude range in which it is observed. It may be seen from Fig. 4.3 that the summer $n_e(h)$ profile is wider, and for this reason larger values are observed in winter than in summer only in the region adjacent to the layer maximum, since the thermosphere is hotter in summer, so that the temperature of the neutral particles and of the plasma in the $F2$-layer is higher. The scale heights are correspondingly larger, especially so in the region above the layer maximum. In the region below the maximum the layer-broadening effect is mainly due the decrease in the zenith distance of the Sun in summer, and to the accelerated rate of ion formation at these altitudes, causing the n_e-values to be higher in summer than in winter.

TABLE 4.2

Quantitative contributions of individual aeronomic parameters to the calculated n_e^m and h_m-values.

$\Delta\lg n_e^m$		Δh_m	
Term in (4.1)	Value	Term in (4.2)	Value
$1.08\Delta\lg[O]$	0.089	$50\Delta\lg[O]$	4.15
$-0.65\Delta\lg\beta$	0.289	$50\Delta\lg\beta$	-22.00
$9\cdot10^{-3}\Delta\omega$	-0.055	$1.55\Delta\omega$	-9.55
$\lg I_1/I_2$	-0.040		
		Overall contribution	-27.4
Overall contribution	0.284	Observed	-26.0
Observed	0.250		

Thus, owing to the higher temperatures (mainly above the layer maximum) and to an enhanced electron concentration due to ion formation (below the layer maximum) in summer time, the summer layer is broader than the winter layer. There is also an additional dynamic factor, which causes compression of the winter layer /274/. It has already been said that the downward wintertime drift produced by thermospheric winds is faster, with consequent diminution of the effective scale height of the plasma distribution above the layer maximum (sec. 2.3.2).

We shall now consider the planetary variation chart of f_oF2 during the year.

We have already mentioned that there is an asymmetry between the Northern and the Southern hemispheres, as manifested by the differences in the amplitudes of annual and semiannual variations of f_oF2 components in the two hemispheres /97/. The extent of the manifestation of this symmetry also depends on the solar activity level /204, 232, 289/.

Asymmetry is usually attributed to the difference between the ratio of the geographic and geomagnetic coordinates in the two hemispheres. However, the f_oF2 asymmetry is still evident for stations in the two hemispheres with similar values of φ, ϕ and I /235/.

As was seen in sec. 3.2, the MSIS model yields asymmetric neutral compositions of the two hemispheres when used in the calculations. We shall see how the observed asymmetry of the $F2$-region in the northern and southern hemispheres can be interpreted in terms of the proposed computation routine.

In a similar way to Mikhailov and Boenkova /52/, we shall compare two stations: Washington, D.C. (U.S.A.) ($\varphi = 38.7°N$, $\lambda = 77.1°W$, $I = +71°$, $D = -7°$) in the northern hemisphere and Watheroo (Australia) ($\varphi = 30.3°S$, $\lambda = 115.9°E$, $I = -65°$, $D = -2°$) in the southern. These stations are located at similar geographical latitudes and have similar magnetic dip and inclination angles, so that this comparison may reveal reasons for the asymmetry which is not due to differences in the coordinates. We shall choose a time close to midday (14 hrs) under magnetically quiet conditions ($A_p = 4$), at three solar activity levels $F_{10.7} - 100$, 150 and 200 and compare the calculated with the experimental f_oF2 values by using data taken from /73/ for W-values of 50, 100 and 150, which correspond to the solar activity levels chosen. We shall consider only one $F2$-layer parameter, the critical frequency f_oF2 since only this parameter yields sufficiently reliable information about the planetary distribution measured at a global network of ionospheric sounding stations (Fig. 4.4).

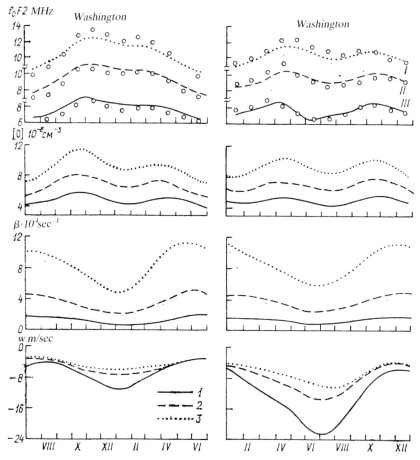

Fig. 4.4. Observed (circles) /73/ and calculated (1–3) annual variations of f_oF2, and the values of [O], β and w at the stations Washington, D.C. and Watheroo, Australia, at $F_{10.7}$ of 100 (1), 150 (2) and 200 (3), used in the calculations.

It is seen from the figure that, on the whole, calculated f_oF2-values give an accurate description of the observed variations of f_oF2, and hemisphere asymmetry. In this case the asymmetry is manifested in the fact that in the Southern hemisphere where the equinoctial maxima are distinctive at all levels of solar activity, the winter and summer values of f_oF2 are approximately equal. In the Northern hemisphere, on the other hand, the equinoctial maxima are indistinct, while the winter f_oF2-values are much higher than the summer values, especially during periods of high solar activity. The calculation also reproduces, quite faithfully, the extent of the seasonal anomaly. Thus, in the northern hemisphere, the anamoly $\Delta f_oF2 = (f_oF2)_{winter} - (f_oF2)_{summer}$ is noted at all levels of solar activity, its values at $F_{10.7}$ of 100, 150 and 200 being 1.4, 2.6 and 2.9 MHz, respectively. In the southern hemisphere the seasonal anomaly is only encountered at $F_{10.7} = 150$ and $F_{10.7} = 200$, and the values of $\Delta f_o F2$ itself are lower -0.4 and 1.0 MHz, respectively. At $F_{10.7} = 100$ no seasonal anomaly is noted in the southern hemisphere.

Figure 4.4 shows the annual variations in the concentration of atomic oxygen [O], the loss coefficient β and the vertical drift rate w, at the altitude of 300 km at three levels of solar activity. It is apparent that all parameter values are different at each station, which must be attributed to the asymmetries in the atmospheric models for the two hemispheres, since the geographical latitudes of the two stations are similar.

Let us now consider the quantitative contributions of individual aeronomic parameters to the observed seasonal changes of n_e^m. For this purpose we shall use the relationship (2.45), between the electron concentration in the F2-layer maximum and the values of [O], β and w at the altitude of 300 km, in the form of finite differences.

Table 4.3 shows the seasonal variations for the Washington and Watheroo stations (winter minus summer) of the prinicpal aeronomic parameters at three levels of solar activity, calculated using the MSIS model.

TABLE 4.3

Seasonal variations of Δ (winter – summer) aeronomic parameters.

$F_{10.7}$	Station	Δ lg [O]	Δ lg β	Δw
100	Washington	0.07	−0.33	−5.78
	Watheroo	−0.01	−0.24	−15.10
150	Washington	0.10	−0.32	−3.91
	Watheroo	0.01	−0.25	−8.60
200	Washington	0.11	−0.35	−2.30
	Watheroo	0.04	−0.25	−5.94

Table 4.4 represents a quantitative estimate of the contribution of the individual parameters to the seasonal variations of electron concentration at the F2-layer maximum at various $F_{10.7}$-values.

TABLE 4.4

Quantitative contribution of individual aeronomic parameters to the calculated seasonal variations of n_e^m according to (4.1), at different $F_{10.7}$-values.

$F_{10.7}$	Station	1.08Δ lg [O]	-0.65Δ lg β	$9\cdot10^{-3}\Delta\omega$	Total
100	Washington	0.07	0.21	−0.05	0.23
	Watheroo	−0.01	0.16	−0.13	0.02
150	Washington	0.11	0.21	−0.04	0.28
	Watheroo	0.01	0.16	−0.08	0.09
200	Washington	0.12	0.23	−0.02	0.33
	Watheroo	0.04	0.16	−0.05	0.15

It may be seen from Table 4.3 that, as the activity increases, the amplitude of seasonal variations of [O] increases both in the Northern and in the Southern hemispheres, while the seasonal changes of β remain constant. Table 4.4 shows that the sum total of the seasonal variations of [O] and β enhances the seasonal differences of n_c^m, while the drift attentuates these differences, i.e., has a "negative" effect on the seasonal anomaly of the F2-region, as has already been shown. This is particularly pronounced in the Southern hemisphere at the minimum of solar activity, when the seasonal anomaly is almost negligible.

Thus, the seasonal anomaly is more distinctly manifested at the maximum of the solar activity in both hemispheres; this is related, first and foremost, to the increase in the amplitude of seasonal variations of [O]. This has also been shown by Mikhailov and Serebryakov /56/, who calculated the F2-region using the OGO–6 model. It may be seen from Table 4.4 that the increase in activity is also accompanied by a decrease in the negative contribution of the drift to $\Delta\lg n_c^m$, particularly in the southern hemisphere.

In the northern hemisphere the seasonal anomaly is more pronounced than in the southern hemisphere. This is mainly due to the fact that the amplitude of variation of the concentration of atomic oxygen at the altitudes of the F2-region is larger in the northern than in the southern hemisphere. At the minimum of solar activity the negative effect of the drifts is particularly pronounced. As a result, the seasonal anomaly is absent in the southern hemisphere at minimum solar activity, while it persists in the northern hemisphere. In the southern hemisphere the increase in the solar activity level is accompanied by an increase in the seasonal variations of [O] and a marked decrease in the compensating effect of the drifts, resulting in a seasonal anomaly in the southern hemisphere as well. At the maximum activity level (Table 4.4) the effects of β and the drifts are about the same in both hemispheres. However, the contribution of [O] to $\Delta\lg n_c^m$ in the southern hemisphere is small, and the seasonal anomaly is therefore larger in the northern hemisphere.

In addition, at all activity levels, distinct semiannual variations of n_c^m are observed in the southern hemisphere. These are mainly annual in the northern hemisphere; the reason for this can also be put down to the difference in the seasonal variations, since the annual variations of n_c^m result from the superposition of the semiannual component, related to the variations in the atomic oxygen concentration, and the annual component, with an amplitude depending on [O], β and w. Since the annual component of the seasonal anomaly is small in the southern hemisphere, the semiannual variations predominate there. In the northern hemisphere the annual component is large, and when added to the semiannual component, effectively masks the equinoctial maxima.

Mikhailov and Serebryakov /56/ discussed the mechanism of superposition of annual and semiannual components in order to explain the latitudinal variations in the annual evolution of critical frequencies, and the effect of convergence of equinoctial maxima with the variation of the latitude and of the level of solar activity. The effect of convergence of equinoctial maxima may be described as follows. According to data obtained by ionospheric sounding, the annual evolution of f_oF2 usually contains two equinoctial maxima; as the latitude increases, the time at which these maxima appear shifts towards the winter solstice, this effect being stronger at high levels of solar activity. At low latitudes and minimum activity, these maxima are observed around the equinox. The seasonal anomaly of n_c^m, which is more pronounced at moderate latitudes and at maximum solar activity, may result /201/ from partial superposition of these maxima. Consequently, the explanation of the variation of seasonal anomaly of critical frequencies of the *F*2-region with latitude and solar activity level requires the understanding of the maximum convergence mechanism during the annual evolution of f_oF2.

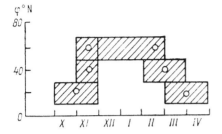

Fig. 4.5. Time of appearance of the maxima in the annual evolution of f_oF2 for three latitudes. Shaded areas indicate the experimental data /201/; circles indicate the calculated values.

We shall consider the data obtained for three latitudes (20°, 40° and 60°) in the Northern hemisphere, under magnetically quiet conditions ($A_p = 4$), at a high solar activity level ($F_{10.7} = 150$), and at 14 hrs local time. The annual variations of the critical frequencies were calculated under these conditions. Figure 4.5 shows the months (shaded areas) which, according to King and Smith /201/, correspond to the annual maxima of f_oF2. It may be seen that the calculation represents a description of the convergence of the maxima in the annual formation of f_oF2 when the latitude may vary.

As mentioned above, the role played by vertical drifts in the Northern hemisphere is small, while the annual variations of n_e^m are mainly caused by a change in the concentration of atomic oxygen and in the loss coefficient. Figure 4.6 shows the calculated annual variations of $\lg n_e^m$, together with the values of $\lg [O]$ and $\lg \beta$ at 300 km altitude for three latitudes.

We shall now consider the latitudinal variations of these parameters. The location of the maxima in the annual evolution of [O] at different latitudes is practically unchanged, while the amplitude of the seasonal variations increases with latitude. Thus, the January-to-July ratio increases by 80% between the latitudes of 20° and 60°, while the seasonal variations of β are more than 400%. The combined effect of these factors enhances the seasonal component of the annual evolution of n_e^m, and the winter-to-summer ratio of the n_e^m values increases from 1.45 at $\varphi = 20°$ to 2.5 at $\varphi = 60°$. The major contribution to the increase in the seasonal component of n_e^m is due to the seasonal variations of β, as pointed out above.

At low latitudes, where the seasonal variations of β are moderate, the semiannual variations of the concentration of atomic oxygen determine the annual evolution of n_e^m and, as a result, the maxima appear around the equinoxes. As the latitude increases, the amplitudes of the seasonal variations of [O] and β increase with a resulting increase in the seasonal component of the variation of n_e^m. As a result of the superposition of the semiannual component (due to [O]) and the annual component (mainly due to β), the overall maxima in the variations of n_e^m are shifted towards the winter months. The fact that this effect is stronger at the maximum of the solar activity is explained, in accordance with Table 4.4, by the increase of the annual component of the variations of n_e^m.

Thus, it is chiefly the specific variations of the neutral composition, as reflected in modern thermospheric models, which determine the characteristic features of the annual variations of n_e^m – in particular, the seasonal anomaly and its variation with the latitude and with the level of solar activity, and the presence of semiannual maxima and their dynamics.

This approach will now be used in the discussion of the so-called December anomaly.

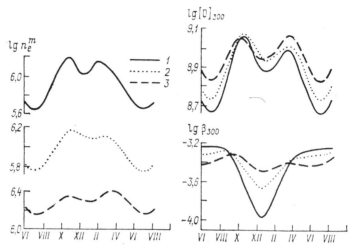

Fig. 4.6. Calculated annual variations of lg n_e^m for three latitudes at enhanced solar activity ($F_{10.7} = 150$), and annual variations of lg [O] and lg β at 300 km altitude.
1) φ = 60°, 2) φ = 40°, 3) φ = 20°.

4.1.2 December Anomaly. Observations of $F2$-layer critical frequencies indicate /64, 97, 233, 289/ that the average global f_oF2-values are higher close to the December solstice than close to the June solstice. Spectral analysis of the annual data for f_oF2 /289, 306/ showed that, in addition to the seasonal component, which is related to the seasonal anomaly of the $F2$-region, there is a non-seasonal annual component, which reaches a maximum in November–January, irrespective of the level of solar activity. With respect to their amplitudes at the various phases of solar cycle, the two components are more or less equal, or the nonseasonal component, which determines the December anomaly, is higher than the seasonal one.

Many hypotheses have been advanced /233, 302/ to explain the December anomaly, all involving the introduction of some supplementary ionization source.

It has been suggested /65/ that the observed 20% average global increase in the electron concentration in December as compared with June may be related to the increased solar flux due to the variation of the distnace between the Sun and the Earth. It is clear, however, that a 6% change in the solar flux is insufficient to explain the effect. On the other hand, an explanation of the December anomaly involves a comparison between the two hemispheres, which have been seen to be asymmetric with regards to the neutral compositions and velocities of thermospheric winds. This problem clearly requires further discussion.

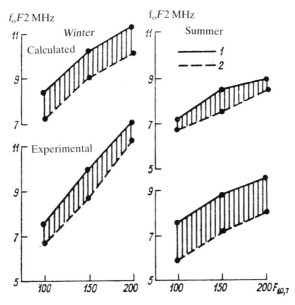

Fig. 4.7. Calculated and experimental /73/ variations of f_oF2
at various solar activity levels at Washington, D.C. and
Watheroo (Australia) stations.
1 – at December solstice; 2 – at June solstice.

In analogy with /6/, we shall compute the June and December f_oF2-values at Washington and Watheroo stations, which are located in different hemispheres and initially ignore the 6% difference in the incident flux, I_o. We will choose three levels of $F_{10.7}$ – 100, 150 and 200 – 14 hrs local time, and magnetically quiet conditions.

Figure 4.7 shows the calculated and the observed /73/ values of f_oF2 for the December and June solstices. The comparison has been made for local winters and summers. It is clear that, for both seasons, the calculations agree with the experiments, and thus reflect the fact that December f_oF2-values are higher. In addition, according to the observations, the effect of December anomaly does not much vary with the solar activity and therefore for all $F_{10.7}$-values studied, those in December exceeded the June values by about 1.0 – 1.5 MHz.

In order to explain the mechanism of this effect we shall consider the variations of aeronomic parameters in the same way. Figure 4.8 shows the variations of the temperature, the concentration of atomic oxygen, the loss coefficient and the vertical drift rate at 300 km altitude, at the December and June solstices, for three levels of solar activity. We may note that in winter the southern hemisphere is warmer than the northern hemisphere while in the summer the temperatures of the two hemispheres are roughly equal. This is reflected in the variation of β. However, the concentration of atomic oxygen is independent of the season, and is higher in December than June.

A similar argument involving the finite differences equation (4.1) shows that the principal contribution to the December increase in the n_c^m-value during local winter is from the loss coefficient and drifts. These effects cause the n_c^m-values in the northern hemisphere – i.e., the December values of n_c^m – to increase, since the absolute values of the rate of the vertical drift and the β-value in the northern hemisphere are lower than in the southern hemisphere (Fig. 4.8).

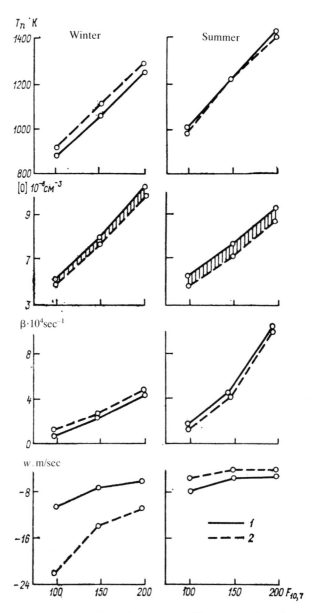

Fig. 4.8. Variations of temperature T_n, concentration of atomic oxygen [O], loss coefficient β, and the rate of vertical drift w, at 300 km altitude, for various solar activity levels. 1 – for December solstice; 2 – for June solstice.

During the local summer the principal contribution to n_e^m variation is made by atomic oxygen, while the contributions of β and w are of secondary importance. Thus, during the local summer, the fact that the December values of n_e^m are higher than the June values is due to the higher concentration of atomic oxygen in the southern (winter solstice) as compared to the northern hemisphere (summer solstice).

Thus, the effect of December anomaly, i.e., the fact that the f_oF2-values are higher in December than in June, may be explained in terms of seasonal changes of aeronomic parameters, without introducing supplementary sources of ionization, except for the UV solar radiation. The 6% increase in the integral solar flux in December–January, due to the change in the distance between the Earth and the Sun, may be considered in calculations, and will enhance the difference between the December and the June values, in agreement with experimental findings. (Fig. 4.7).

Therefore it may be concluded that the principal cause of the December anomaly, noted in the *F2*-layer, is the asymmetry between the hemispheres with respect to the composition of the neutral atmosphere and the rates of thermospheric winds.

The question of the asymmetry between the two hemispheres is a separate, highly interesting problem. According to the experimental data obtained in the course of the past few years, many aeronomic parameters display the North–South asymmetry /296/. According to /238, 239/, there are considerable differences in the seasonal variations of the concentration of helium in the two hemispheres – by a factor of 9 in the northern and by a factor of 16 in the southern hemisphere. It was suggested, in this context /94/, that the upper atmosphere seems to be hotter in the southern than in the northern hemisphere. This may partly be due to the fact that in December–January the Earth is closer to the Sun, as well as to the asymmetry of the geomagnetic field /209/. Differences between the temperatures of the upper troposphere in the two hemispheres have been noted /205/. Clearly, all these factors may influence the effect of the turbulent conditions at the turbopause level, which would directly alter the concentrations of atomic and molecular oxygen in the upper atmosphere /2, 3/. In fact, the insignificant seasonal variations in the concentrations of atomic and molecular oxygen, and the presence of reverse semiannual variations of O_2, given by the MSIS model in the southern hemisphere, seem to confirm the hypothesis of the different effects of turbulence in the two hemispheres.

4.2 DIURNAL VARIATIONS

In the analysis of annual variations the only conditions considered were those around midday in the *F2*-region, and the relative role played by individual processes in the formation of this region remained more or less the same throughout the year. Furthermore, under around-noontime conditions the *F2*-region may be considered to be approximately stationary, since at that time of the day the parameters of the neutral atmosphere and the zenith angle of the Sun are both near their extreme values, and their variations are accordingly small.

In considering the diurnal variations in the *F2*-region it should be borne in mind that different processes predominate at different times of the day. Consequently, in the *F2*-region, the description of the diurnal variations is more complex than of annual variations.

Data obtained at the Millstone Hill station indicate that the diurnal variations of n_e^m are of two types – winter and summer variations /133, 136, 137, 143, 146/.

The following are the typical features of the winter-time variations:

1) The major diurnal variations of n_e^m (by up to one order of magnitude);

2) The diurnal maximum is reached at about 13 hrs;

3) A drop in the electron concentration during the first few hours following sunset to some background level, which may be maintained throughout the night; and

4) In a number of cases there is a post-midnight increase of n_e^m, which is accompanied by the descent of the layer and a decrease in plasma temperature.

The following are the typical features of summer-time variations:

1) Insignificant diurnal variations, with a day-to-night ratio of about 2:1;

2) A diurnal maximum of n_e^m, which usually occurs in the evening (18 – 20 hrs); and

3) A continuous decrease in the electron concentration throughout the night, until sunrise.

There are also other features which distinguish the summer from the winter type variations. Thus, the summer values of h_m are usually larger than the winter ones by 20 km or more. Moreover, the depth of the F-layer is larger in summer, both above and below the maximum. These seasonal differences have already been discussed in the context of the seasonal anomaly.

Summer variations succeed winter variations, and *vice versa*, around the equinoxes. The change is quite rapid and takes about two weeks. Thus, according to the observations made in 1969 /146/, the transition from the winter to the summer type n_e^m variations took place during the period 25th March – 10th April, while the reverse transition from the summer to the winter variations took place on 14th – 15th October. These transitions indicate a sharp change in the composition of the upper atmosphere which, according to Evans /137–139, 143/, may be connected with seasonal changes of the system of thermospheric winds. These transport atomic oxygen from the summer to the winter hemisphere, thus altering the [O]/M ratio, where M is the concentration of molecular oxygen and nitrogen. An alternative theory /2/ includes the annual changes in the rate of photodissociation of O_2 molecules, along with the seasonal variations in the turbulent diffusion coefficient. This approach makes it possible to explain both the seasonal variations in the concentration of atomic oxygen and the presence of semiannual peaks in its annual cycle.

Figures 4.9 and 4.10 illustrate the diurnal variations of the $F2$-region on magnetically quiet dates (in winter and in summer) according to the results obtained at Millstone Hill station /146/. The typical features of diurnal variations enumerated above are clearly seen in the diagrams, so that the dates chosen may be regarded as typical for the conditions considered.

In studying the factors responsible for the diurnal variations of the parameters of the $F2$ maximum layer, we may first consider the diurnal variations of the altitude h_m. In all seasons, the maximum h_m-values are noted /4, 146/ around midnight, while the minimum values are noted in the morning, the drop in h_m being 100 km or more. The rise of the layer at night is caused by the increase in the upward drift, produced by the southward thermospheric wind (Fig. 4.9). The uplift of the layer (cf. sec.

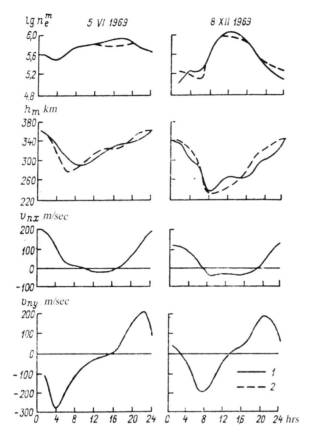

Fig. 4.9. Diurnal variations of observed (1) and calculated (2) values of lg n_e^m and h_m, and calculated velocities of thermospheric winds, v_{nx} and v_{ny}, at 300 km altitude.

2.3.2) may be explained by entrainment of the plasma by the moving neutral atmosphere, and the fact that the plasma is magnetically dominated, so that it can only move along the lines of force of the magnetic field.

Since sunset is followed by a continuous decrease in both the concentration of atomic oxygen and in loss coefficient, the altitude h_m should decrease in accordance with equation (2.47). However, during the first half of the night h_m in fact increases; a decrease only occurs during the second half of the night, and is an indication of the variation in the vertical drift rate, w (cf. Fig. 4.10). Thus, on 8th December 1969, between 18 and 24 hrs, the sum (lg [O] + lg β) decreased by about 0.3, while the vertical drift increased to about 40 m/sec, with the resulting increase of about 25 km in the altitude h_m, in agreement with results of the calculation (cf. curve 2, Fig. 4.9). Were it not for the nighttime vertical drift, the altitude of the layer maximum would decrease continuously throughout the night.

During the transition period from nighttime to daytime conditions, h_m rapidly decreases for various reasons. At sunrise, with the beginning of intensive ionization

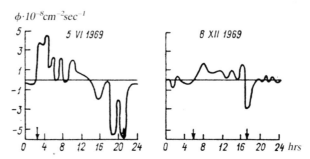

Fig. 4.10. Variation of the ionosphere-protonosphere flux, ϕ, at 650 km altitude. Arrows show the moments of sunrise and sunset at 200 km altitude /164/.

of the lower $F2$-region, the daytime formation mechanism begins to be operative, when the layer maximum shift downwards compared to its nighttime altitude /65/. It was shown by Gliddon and Kendall /166/ that this effect occurs even if the temperature and the atmospheric composition remain constant throughout the day. Under real conditions, the atmosphere remains sufficiently cool during the morning hours and, under otherwise identical conditions, the layer maximum is situated at a lower altitude than in the afternoon, for example, when the atmospheric temperature is higher.

Another factor, which is responsible for the downward motion of the layer, is the slower upward drift, which is mainly due to the decrease in the velocity of the meridional component of the thermospheric wind v_{nx}. At the Millstone Hill station this effect is more pronouned due to the negative zonal component, v_{ny}, which also produces a downward drift, as the magnetic inclination angle is negative ($D = -15°$) at this site.

In daytime, as the temperature increases, the concentration of atomic oxygen and the loss coefficient, which is proportional to the concentration of molecular components, both increase in accordance with (2.44), resulting in a higher altitude of the layer maximum, h_m. Accordingly, the increase of h_m from morning to afternoon is due to the increase in the temperature of the atmosphere.

In the evening, as the rate of ion formation decreases, a transitional period follows, during which the altitude of the layer maximum increases to its nighttime value. This process is enhanced by the change in the direction of the thermospheric winds (Fig. 4.9), since an upward drift causes an additional uplift of the layer. The drift rate increases towards midnight, and the maximum layer altitude h_m is noted around midnight, as explained above. Thus, thermospheric winds play a significant part in the diurnal variations of the altitude of the $F2$-maximum layer. If the winds are not taken into account, the diurnal variations in h_m are only 40–50 km /108, 203, 240, 274/, which is less than one-half of the experimental values. If the diurnal variations of atmospheric temperature and consequent variations in [O] and β are allowed for, and there is no vertical drift, it may be seen from (2.44) and (2.47) that the h_m-values obtained at 12 hrs and 24 hrs on 8th December 1969, for example, will be 282 and 294 km, respectively.

As shown in sec. 2.3, the vertical drift also markedly affects the electron concentration, n_e^m. Winds are responsible for a number of special features in the variation of the electron concentration in the *F*2-region. These include the evening time increase of n_e^m in summer, and the formation of the nighttime *F*2-region.

4.2.1 The Nighttime *F*2-Region. The principal problem involved in the study of the nighttime *F*2-region is how to explain its formation mechanism. It is known that, despite the absence of any ionization source, the electron concentration n_e^m never drops below a certain background level, even during the long winter nights; this level is 10^5 cm^{-3} /303/. Variations of n_e^m, and even an increase may, also be observed at night. Data on the total electron content, N /121/, collected at the mid-latitude Stanford station during 313 nights between October 1964 and December 1965, indicate that the N-value remained constant throughout the night in 44%, decreased in 30%, increased in 10% of the cases, and underwent irregular variations on the remaining nights. However, this general result does not take into account seasonal differences typical of the *F*2-region. Thus, according to /146/, the decrease in n_e^m is observed mainly during the summer nights, while during the winter nights it is observed only during the first few hours after sunset, n_e^m remaining unchanged thereafter, or even occasionally increasing after midnight /134, 137, 146/. This type of nighttime evolution of n_e^m at mid-latitudes also results from statistical processing of the monthly median values of critical frequencies, obtained by the world-wide network of ionospheric sounding stations /73/.

A more general approach to the behavior of the *F*2-region at night is to attempt to explain formation mechanism of the nighttime ionosphere. One possible mechanism /163, 165, 173, 303/ is the influx of ionization from the protonosphere which, according to different workers, varies between $4.7 \cdot 10^7$ cm$^{-2} \cdot$sec^{-1} /303/ and $2 \cdot 10^9$ cm$^{-2} \cdot$sec^{-1} /275/. The wide range of flux values is due to the differences in the rates of recombination processes postulated by the different workers.

According to modern estimates /147/, the nighttime protonospheric flux at the Millstone Hill station is between 10^7 and 10^8 cm$^{-2} \cdot$sec^{-1}, the most probable value being $3 \cdot 10^7$ cm$^{-2} \cdot$sec^{-1}. However, such small flux values cannot explain the observed electron concentrations or the altitude of the nighttime *F*2-layer at the postulated recombination rates, so that another mechanism must be sought. It was suggested /115, 137, 242, 244, 272/ that the thermospheric winds may constitute such a mechanism since they are directed towards the Equator at night, and cause uplift of the layer and a reduced recombination rate. Based on this theory, small plasma fluxes from the protonosphere would be sufficient to maintain the electron concentration at a relatively high level /243/. A specific calculation, illustrating this mechanism, will be performed below.

We shall now consider this problem in some detail, starting with the non-stationary behavior of nighttime *F*2-region. According to observations /146/, small temporal variations in the electron concentration n_e^m and in the plasma temperature are noted during partial nighttime periods. It has been concluded (see, for example, /163, 183/) that a stationary description of the *F*2-region is suitable for these periods. However, the nighttime *F*2-region is inherently a non-stationary formation. Thus, while under conditions prevailing around noon the characteristic time variation τ of electron concentration with respect to recombination (1/β) is about 1½ hours in the layer maximum, while around midnight it is about 10 hours. This means that during the day the variations of parameters such as q and β are "watched" to a considerable extent.

The nighttime situation is different. Since the characteristic time τ is longer, the electron concentration at any given moment of time largely depends on its previous history. For this reason, as shown in /54/, if erroneous values are assumed for the initial distribution of n_e^m in the evening, this will affect the results obtained during the entire night. Furthermore, all these processes take place against the background of a continuously cooling and contracting thermosphere and, even if the electron concentration in the layer maximum remains constant, it still decreases with time near the upper boundary of the region (600 km) /146/. Accordingly, a description of the F2-region necessarily involves the solution of a nonstationary diffusion equation, in which the initial distribution is given. It may also be seen from Fig. 4.11 that nonstationary conditions should be taken into consideration. Under stationary conditions, the flux value above the layer maximum is practically constant, due to the low effectiveness of the recombination process. In order to explain the observed n_e^m-values at the upper boundary, the postulated fluxes would have to be almost 10 times larger than the observed ones. In a nonstationary case, the flux profile would have an altogether different shape and, at the upper boundary, its value would agree with the observations.

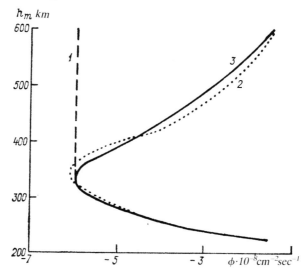

Fig. 4.11. Profiles of the plasma flux, calculated for stationary conditions (1), nonstationary conditions (2), and the
experimental values /146/ (3).

We shall consider the results of calculations performed on four magnetically quiet days, in different seasons, and compare them with observations made at the Millstone Hill station. Figure 4.12 /54/ shows the temporal variations of n_e^m, h_m and w. As in sec. 3.4, the rate of the vertical drift w, was chosen so as to obtain agreement between the calculated and the observed h_m-values. The n_e^m-values, calculated for this drift rate, were compared with the experimental values. Fig. 4.12 shows the vertical drift rates obtained in this way, as well as the experimental drift rates obtained for one of the above dates /140/. The initial situation assumed in the calculations was the experimental electron concentration profile (at 20 hrs), while a constant downward plasma flux of 10^8 cm^{-2}·sec^{-1} was assumed to be present at the upper boundary. It may be seen from Fig. 4.12 that if we ensure that the h_m-value employed is correct by a suitable choice of value for the drift rate w, we also obtain the true value of n_e^m. This confirms the importance of winds in any study of the nighttime *F*2-region.

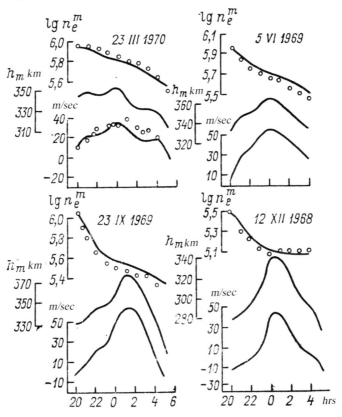

Fig. 4.12. Variations of electron concentration lg n_e^m, during the night (points represent the experimental data), the observed values of the altitude of the maximum h_m, and the calculated vertical drift rates, w.

These calculations reflect the aforementioned special features of the variation of electron concentration at nighttime. Thus, n_e^m decreases during the first few hours after sunset, and continues to decrease until sunrise in all seasons except in winter, when the decrease ceases after midnight, n_e^m then remaining constant at some background value. Let us consider the physical reasons for these differences; in order to do this, we shall compare the results obtained on a winter day (12 December 1968) and on a summer day (5th June 1969).

During the first few hours after sunset the main process determining the temporal variation of the electron concentration is recombination. As the layer is lifted by the vertical drift, the rate of recombination decreases, and the effect of the flux from the protonosphere gradually becomes greater. After midnight, the thermosphere continues to cool down and contract, while the $F2$-region is held at high altitudes by thermospheric winds, and the recombination effect progressively weakens. Finally, the rate of plasma influx becomes equal to the loss rate by recombination, and the drop in the electron concentration ceases. These conditions also persist in winter.

Let us follow the effect of this mechanism by taking specific numerical values. In order to do this, we shall compare the total amount of the plasma entering the $F2$-region (by influx) within a definite time period to the total number of recombinations during the same period. We shall take a 4-hour time period (23 hrs to 3 hrs), and perform the calculation for the winter and then for the summer dates. A comparison of the total number of recombinations with the number of particles introduced from the protonosphere by influx across the upper boundary on the winter date gives $1.53 \cdot 10^{12}$ and $1.44 \cdot 10^{12}$ particles, respectively, accounting for the observed constancy of n_e^m. The corresponding figures on the summer date are $2.5 \cdot 10^{12}$ and $1.44 \cdot 10^{12}$ particles, respectively. It is apparent, therefore, that in summer the number of charged particles undergoing recombination exceeds those entering the region by plasma influx from above. In this sense the summer night resembles the first few hours after sunset during the winter, when recombination constitutes the main process.

As mentioned before on winter nights the decrease in electron concentration to the background level is accompanied by a typical increase of n_e^m after midnight. This effect can be followed by studying the variation of n_e^m, and the data on the total electron content /93, 98, 121, 133, 134, 146, 193, 287, 290/. Observations carried out at Millstone Hill /145, 146, 148/ indicate that the increase in n_e^m begins soon after midnight. The maximum is reached at about 2 hrs (± 2 hours), when the increase in the electron concentration is usually accompanied by a decrease in the altitude of the maximum. The increase in n_e^m is characteristic of winter nights only, but is not always in fact recorded. According to /134, 146, 290/, the nighttime maximum of n_e^m is observed both at maximum and at the minimum of solar activity in the range of geomagnetic latitudes of 30–55°, and the probability of its appearance decreases on moving both towards higher and lower latitudes /290/.

In order to clarify the reasons for the increase in electron concentration, we shall consider (as was done by Mikhailov and Ostrovskii /55/) the night of 25/26 November 1968 /145/, when this effect was particularly sharply manifested. Figure 4.13 shows the observed variations of n_e^m and h_m. It follows from the graph that n_e^m rapidly decreases during the first few hours after sunset, after which the decreases is resumed.

The thermospheric MSIS model was used in the calculations, but the evolution of the temperature T_n during the night was corrected in accordance with the observed variations in the ion temperature at altitudes of 200–250 km which is known /91/ to reflect the temperature of the neutral particles undre these conditions. Thus the calculation in fact reproduces the main features of the variation of n_e^m during the night.

In the same way as above, it may again be shown that between 23 hrs and 2 hrs –
a period corresponding to an increase of n_e^m – the influx across the upper boundary
into the *F*2-region was $1.08 \cdot 10^{12}$ particles, while $4.75 \cdot 10^{11}$ particles were lost through
recombination. Thus, by 2 hrs the balance is positive. A similar comparison made for
the period between 2 hrs and 5 hrs shows a loss of $5 \cdot 10^{11}$ particles.

It should now be clear that, depending on the ratio between the influx and
recombination of the plasma, n_e^m may vary during the night in different ways: a
continuous decrease or an increase during some periods, or the electron concentration may remain unchanged.

The comparison of two winter-time dates given in Figs. 4.12 and 4.13 shows that
the different types of n_e^m-variation during the night are due to the temperature
changes in the neutral atmosphere. Thus, the November day had a low temperature
of $T_n \approx 700°K$ which fell to about 600°K at about 2 hrs, while on the December date
high ion temperatures T_i, were recorded, and the MSIS values for the temperature of
the neutral paticles were used in the calculations. If the atmosphere is cooler, the
values of the loss coefficient at a given altitude are lower, favoring plasma
accumulation, as was in fact the case on the November date.

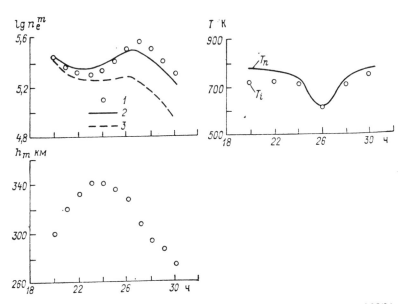

Fig. 4.13. The winter increase in the electron concentration n_e^m on the night of 25/26
November 1968 and the experimental variations in the h_m, ion temperature T_i and the
temperature T_n of the neutral particles, used in the calculations.
1 – Experimental data /145/; 2 – calculated for constant flux at the upper boundary (10^8
$cm^{-2} \cdot sec^{-1}$); 3 – calculated for zero flux.

Analysis of the results of the observations of n_e^m and T_i, made at the Millstone Hill
station during 1968–1970, shows that the cooling of the winter thermosphere, with
minima in the T_i-values, is a fairly typical phenomenon for the nighttime *F*2-region
in winter, but in some cases there is no significant rise in the electron concentration.
This indicates that the mechanism of n_e^m increase is due not only to the cooling of the

thermosphere, but also to the magnitude of the protonospheric flux. To illustrate the role played by the flux, Fig. 4.13 shows the calculated variations of n_e^m (curve 3) under the same conditions as in the main calculation (curve 2), except that the flux at the upper boundary is assumed to be zero. It is seen that in such a case n_e^m does not increase at all; this is understandable, since no complementary plasma flows into the region under consideration.

Thus, the increase, decrease or constancy at the background level of electron concentration during the night depends on the ratio between the influx of the plasma from the protonosphere, which is the source of the plasma present in the $F2$-region and its recombination rate, which is ultimately determined by the temperature of the neutral atmosphere. These cases may be described by assuming small flux values ($\phi \leq 10^8 cm^{-2} \cdot sec^{-1}$) which agree with the observed values, while the nature of the variation of n_e^m during the course of the night depends on the atmospheric temperature.

It should be noted that the protonospheric flux undergoes irregular variations during the night, and the case of a small increase of electron concentration /146, 148/ may be attributed to individual surges of flux, approximately up to $2 \cdot 10^8 cm^{-2} sec^{-1}$. However, if the increase of the electron concentration n_e^m is larger (i.e., if it is approximately doubled, as was in fact the case on the night of 25th/26th November 1968 considered here), the fluxes would have to be much larger (up to $6 \cdot 10^8 cm^{-2} \cdot sec^{-1}$) than those in fact observed. Therefore, the only possible explanation for such effects is the cooling of the atmosphere.

It follows from the discussion above that an interpretation of the individual features of behavior of nighttime $F2$-region requires detailed analysis of all the conditions prevailing on the day in question. On the other hand, the nighttime increase of electron concentration in the $F2$-region is a fairly typical effect, which is reflected in the median values of n_e^m, quoted in /73/. We may therefore conclude that the nighttime T_n-value given by the MSIS model is high.

Let us consider next the absence of the seasonal anomaly of n_e^m during the night – which is closely connected with the problem of nighttime $F2$-layer formation. In fact, it is seen from Fig. 4.9 that the winter daytime electron concentration is considerably higher than the n_e^m values in summer (seasonal anomaly), while being below the summer nighttime n_e^m values. It was shown in para. 4.1.1 that the main reason for the seasonal anomaly of n_e^m are variations of the neutral composition, as indicated by the $[O]/[N_2]$ ratio, which is lower in summer than in winter. However, this is also true for the nighttime $F2$-region. Moreover, the nighttime seasonal variations of that ratio at 300 km altitude are even more marked, but in spite of this nighttime seasonal anomaly of n_e^m is not observed. It is known (Chapter 2) that the concentration of atomic oxygen determines the rate of ion production ($q \propto I_o [O]$) at the altitudes of the $F2$-region, while the molecular concentrations (mainly N_2-concentration) determine the value of the loss coefficient β. The source of photo-ionization of atomic oxygen is absent in the nighttime $F2$-region, while the $F2$-layer itself is raised above the zone of high recombination rates by the thermospheric winds (in summer and in winter) and as a result, the coefficient β in the layer maximum around midnight is about 10^{-5} sec^{-1}, which is less than the daytime value by one order of magnitude. It follows that the seasonal differences in the neutral composition are not responsible for the observed differences in n_e^m at night, which must therefore be due to other causes.

Figure 4.9 shows that the n_e^m-values are the same at sunset (at 17:30 hrs in winter and at 21 hrs in summer), while the altitude h_m is higher by about 50 km in summer than in winter, since on a summer day the meridional and zonal components of the thermospheric wind, v_{nx} and v_{ny} respectively, change direction long before sunset, and the strong ($w = 52$ m/sec) upward drift lifts the layer, thus reducing the effectiveness of the recombination. In the layer maximum lg $\beta = -4.4$ at 21 hrs. On a winter day (cf. Fig. 4.9), during the sunset period, the thermospheric winds just change directions, their velocities are low, and the velocity of the vertical drift is also low ($w = 2$ m/sec). Thus, the layer remains in the region of relatively high recombination rates (lg $\beta = -3.9$ in the layer maximum). In the first few hours after sunset, when the drift is still weak, there is a large drop in the electron concentration. This may be seen in Fig. 4.9 which shows that during the four hours following sunset the lg n_e^m-value decreased by 0.2 in summer and by 0.4 in winter. Thus, the winter n_e^m at night is lower than its summertime value, in contrast to the seasonal anomaly displayed for this parameter in daytime.

4.2.2 Calculation of daytime variations of n_e^m and h_m. We shall consider the diurnal variations of the parameters of the layer maximum calculated for 32 magnetically quiet dates during 1969–1970, at the Millstone Hill station /146, 148/. The data were obtained during different seasons and at different levels of solar and geomagnetic ($A_p \leq 15$) activity. The computation routine used has already been described in /57/ and explained in sections 2.1 and 2.2 above. The comparison of calculated with experimental data was performed at four moments of time (0 hrs, 6 hrs, 12 hrs and 18 hrs), and the results are shown in Fig. 4.14.

It is apparent from Fig. 4.14 that at the times stated the electron concentrations lg n_e^m and altitudes h_m, were usually within the above mentioned range (± 0.15 for the electron concentration lg n_e^m, and ± 20 km for the altitude, h_m, of the layer maximum). At sunset (18 hrs), the calculated h_m-values are higher by about 20 km, while the calculated lg n^m-values are usually lower than the experimental ones. Several workers /130, 131, 132, 135, 203, 242, 272, 274, 283, 284/. have given various explanations for the mechanism of the increase in the electron concentration in the *F*2-region, based on different principles.

It follows from these studies that the observed variations of the parameters of the *F*2-layer maximum around sunset are not yet perfectly understood. The same applies to the problem of describing variations of n_e^m during the morning hours (6 – 10 hrs).

It may be seen from the figure, on the other hand, that the calculated values of this parameter for the morning hours tend to be high. This indicates that phase shifts, which are not considered in the MSIS model, play a part in the diurnal variations of the composition and the temperature.

The values calculated for h_m at around midnight are about 15–20 km. This may be due to the fact that the temperatures T_n yielded by the MSIS model are elevated, as was noted in the preceding section. The scatter of the n_e^m-values around midnight illustrates the difficulty of describing the nighttime *F*2-region, in particular the ratio between the protonospheric flux and the temperature of the nighttime atmosphere.

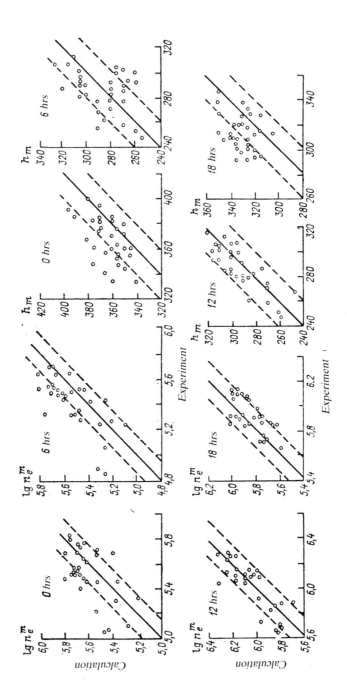

Fig. 4.14. Calculated and observed /146, 148/ values of lg n_e^m and h_m for four moments of time. Broken lines indicate deviations of ±0.15 for lg n_e^m and of ±20 km for h_m.

A successful description of the diurnal variations of the $F2$-region requires a more detailed knowledge of the neutral composition and of the temperature than afforded by the models currently employed. Nevertheless, even today, the parameters of the layer maximum can be computed to within 20–30% using the procedure presented here; this is of potential interest in the switch-over to deterministic prediction of the state of the $F2$-region.

CHAPTER 5. PLANETARY PREDICTIONS

As mentioned in Chapter 1, predictions of the ionosphere may be based on various principles; in particular, different ionospheric models may be used. It is always assumed that under the similar helio-geophysical conditions, the response of the ionosphere will be the same as it has been in the past. The prediction of the state of the ionosphere will then consist of two stages – a prediction of the input parameter values some time ahead, and the calculation of the ionosphere proper, using these values. These two tasks are mutually independent, and the quality of the prediction will depend on the errors involved in each individual stage.

Many input parameter must be known before the ionosphere can be calculated. Thus, the predictive input parameter in the statistical models which will be considered below, viz., "Prediction of MUF" and "Geometric Parameters of the Ionosphere", is the monthly average number of sunspots W. In the semiempirical model (cf. sec. 5.2) the calculation of the theoretical part involves indexes of solar and geomagnetic activities, while the number of sunspots is the input parameter of the statistical part of the model.

In the deterministic approach to the prediction of the ionosphere, the input (predictive) parameters are the indexes of solar ($F_{10.7}$) and geomagnetic (A_p) activities for given day, while the calculation of the ionosphere is carried out according to a theoretical model. The index $F_{10.7}$ characterizes the flux of the solar short-wave radiation, while the neutral composition and the atmospheric temperature also depend on $F_{10.7}$ and on the index A_p (Chap. 3). These parameters can now be more or less reliably predicted 1–3 days ahead, so that the prediction is also valid for this period. With regard to long-term predictions based on this principle, the only parameter that can be calculated in this way is the quiet state of the atmosphere for the monthly average value of $F_{10.7}$.

The second aspect of the prediction problem is the calculation of the ionosphere for the given values of input parameters. Obviously, the more accurately a given computation procedure reflects the observed state of the ionosphere, the more accurate the prediction will be. The accuracy of the description of the ionosphere may be found by epignosis. We shall therefore see in the discussion that follows how accurately the known past states of the ionosphere, i.e., the helio-geophysical conditions, may be described by a number of models. We shall also compare the results obtained using different models with one another.

5.1 STATISTICAL PROBABILITY PREDICTION OF THE F2-REGION

As we have mentioned in Chapter 1, both empirical and statistical models are based on processing long series of observations of ionospheric parameters. By

selecting these data and their statistical processing it is possible to establish their
dependence on geographic coordinates, time of day or level of solar activity. Results
may be presented as charts, graphs, tables, or selected development coefficients. We
do not intend to review all the available empirical models of the ionosphere but will
consider two examples, the "Prediction of MUF" /73/ and "Geometric Parameters of
the Ionospheric F2-Layer" /4/ models, developed by the Institute of Geomagnetism,
Ionosphere and Radiowave Propagation of the USSR Academy of Sciences. These
models, in conjunction with other parameters required to compute the pathways of
the radiowaves, yield the planetary distribution of the monthly median values of both
the critical frequencies of the F2-layer, and the altitudes h_m of the layer maximum.
The input consists of data obtained during 2–3 solar activity cycles from the
world-wide network of ionosphere sounding stations. The values of f_oF2 and of h_m
may be found in the charts or calculated using the analytical relationships involving
the development coefficients of spherical functions /74/.

Analytical description of the planetary distribution of ionospheric parameters is
based on the method of harmonic spherical analysis. According to this method, the
parameters of the ionospheric F2-layer at a given moment of time may be
represented as function of the coordinates $F(X, \Lambda)$ in the form of the series:

$$F(X,\Lambda) = \sum_{m=0}^{M} \sum_{n=m}^{N} [g_n^m \cos m\Lambda + h_n^m \sin m\Lambda] P_n^m (\cos \chi').$$

where Λ is the geomagnetic longitude; X is the modified latitude defined as $X =$
arctan $(I_{300}/\cos \phi)$ where I_{300} is the magnetic dip at 300 km altitude; ϕ is the
geomagnetic latitude; $\chi' = 90 - \chi$; g_n^m and h_n^m are the development coefficients; and P_n^m
is the associated Legendre polynomial of degree n and order m.

The coefficients g_n^m and h_n^m are found by the method of least squares, from the
minimum value of the expression:

$$\sigma = \sum_{i=1}^{I} [f(X_i, \Lambda_i) - F(X_i, \Lambda_i)]^2,$$

where f is the observed value of the parameter of the F2-layer.

The use of geomagnetic coordinates X and Λ instead of the geographic coordinates
simplifies the description of the planetary distribution of the F2-layer parameters, and
makes it possible to reduce the number of terms of the approximation series since the
distribution of the F2-parameters is largely controlled by the geomagnetic field, particularly in the equatorial and polar regions.

The charts and the selected development coefficients are given for even hours of each
month, at several levels of solar activity as determined by the sunspot number W. The
reference levels in the "Prediction of MUF" /73/ model are the four W-values: 10, 50, 100
and 150; whereas in the model /4/ the three W-levels selected are: 10, 100 and 200; the
parameters at intermediate values of activity are obtained by interpolation between the
reference points.

We shall discuss the potentialities of these predictive models, and the accuracy with
which they reproduce experimental data. The main advantage of these models is their
planetary nature, which is important in solving problems in radiowave propagation. Fi-

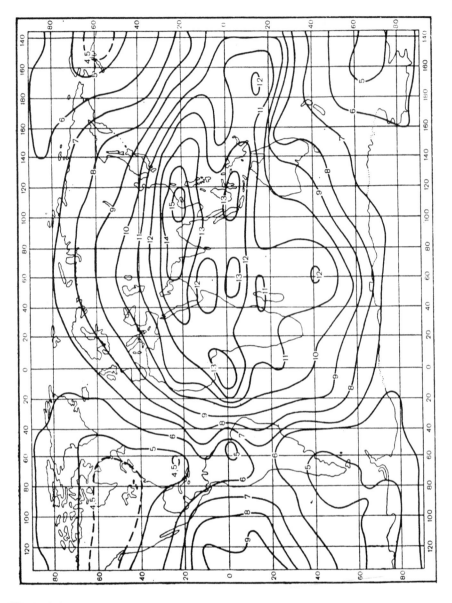

Fig. 5.1. Planetary distribution of the parameter $F2$-0-MUF in September, at 12:00 hrs Moscow official time; W = 100 /73/. Figures marking the iso-lines denote MHz.

gure 5.1, which is taken from /73/, shows the planetary distribution chart of the parameter F2-0-MUF, which is higher than f_oF2 by the value of the gyro-frequency. The charts give the geographical coordinates, for even hours of official Moscow time, for a given level of solar activity. Figure 5.2 shows the chart taken from /4/, which gives the global distribution of h_m as a function of geometric latitude for various local times, for a given level of solar activity. In model /4/ the dependence of h_m on the longitude is taken into account for equatorial latitudes only, in the form of corrections, specified in a certain manner.

A comparison of the predicted critical frequencies f_oF2 at several levels of solar activity with the median values of f_oF2 obtained at different stations showed that the predicted f_oF2-values were accurate to within about 11% /7/.

The accuracy with which the altitudes of the $F2$-layer are represented by the model /4/ is more difficult to estimate, since reliable experimental values of h_m are lacking. However, it is possible to compare the predicted data with those obtained by averaging over several years by the incoherent scatter method. Such a comparison was in fact performed for four seasons at a high solar activity level ($F_{10.7} = 150$, W = 100) (Table 5.1). The averaged h_m-values were those obtained at Millstone Hill station during 1967–1970 at 14 hrs local time /141, 145, 146, 148/.

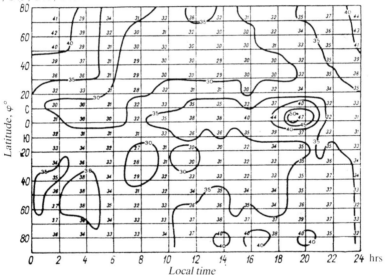

Fig. 5.2 Planetary distribution of h_m (in tens of km) in September (W = 100) /4/

TABLE 5.1

Observed and predicted /4/ values of h for four seasons.

Data	Season			
	winter	spring	summer	autumn
Observed	270	315	310	290
Predicted	285	320	325	300

A comparison of the above data indicates that on the average the daytime model faithfully reproduces the annual variations of h_m, even though the absolute h_m-values are somewhat high. The error is small, and remains within the accuracy of the determinations of individual h_m-values by /4/.

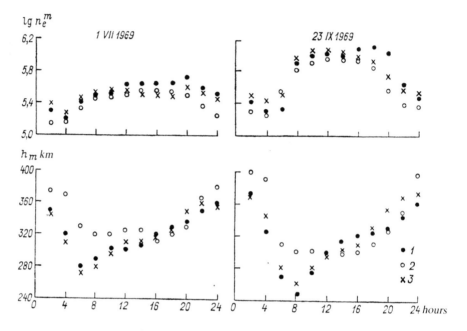

Fig. 5.3. Diurnal variations of lg n_e^m and h_m: (1) observed; (2) calculated by /4, 73/; (3) calculated by the deterministic method.

It is of interest to discuss the applicability of these data to the description of individual days, under a given set of selected helio-geophysical conditions. This is generally not trivial, since the models are constructed for median values of f_oF2 and h_m, and dependence on the level of solar activity is expressed by means of Wolf numbers. It is known that the number os sunspots is only an average index of the intensity of the solar UV radiation, which is responsible for the formation of the ionosphere. Therefore, there is a correlation between the values of f_oF2, h_m and the Wolf numbers, if these are averaged over a sufficiently long period of time (such as one month). However, there is no such correlation on individual days. On the other hand, the 10.7 cm radiation flux of the Sun may be used as an indicator of the short-wave radiation flux (sec. 3.1). Since there is also an average correlation between $F_{10.7}$ and W, it is possible to find the effective sunspot number W corresponding to the $F_{10.7}$-value observed on a given day, and use it to calculate the ionospheric parameters.

Figure 5.3 shows the diurnal variations of lg n_e^m calculated according to model /73/, and of h_m calculated according to model /4/ observed on two specific dates at the Millstone Hill station. The figure also shows the results of the calculation of the parameters of the layer maximum according to the proposed deterministic procedure. It is seen that calculation according to /73/ yields satisfactory results throughout the day, although the deterministic calculation gives better agreement with experiment. Calculation of the altitudes h_m according to model /4/ yields very high values, especially just after midnight and, even more so, in the morning hours (up to 40–60 km), as compared to the observed values. A deterministic computation, on the contrary, is a fairly accurate description of the special features of diurnal variation of h_m.

5.2 CALCULATIONS USING SEMIEMPIRICAL MODELS

In addition to the statisical models, derived by statistical processing of observation data, there are semiempirical models. They are a combination of the theoretical and statistical models, the electron concentration profile obtained by calculation being corrected using empirical data.

The Irkutsk Semiempirical model /63/. Let us consider the principal assumptions underlying this model, developed at the Irkutsk State University. The model describes the spatial and temporal distribution of the electron concentration in the 100–1000 km altitude range, and consists of three parts: the theoretical, the statistical and the correction method.

The theoretical part is a system of three continuity equations for the atomic ions O^+ and H^+, and molecular ions M^+. The atomic ions are considered as being transported by diffusion while the molecular ions are regarded as being in photochemical equilibrium. The rates of ion production by the major components are calculated from the thermospheric JACCHIA–73 model /189/ and the solar UV radiation spectrum /22, 25/. The reactions studied are:

$$H^+ + O \rightleftarrows H + O^+,$$

which is a reversible reaction, with the rate constants $\gamma_1 = 2.6 \cdot 10^{-11} \sqrt{T}$ and $\gamma_2 = 2.9 \cdot 10^{-11} \sqrt{T}$ respectively, and the following two reactions:

$$O^+ + N_2 \rightarrow NO^+ + N, \gamma_3 = 3.6 \cdot 10_{-10} T^{-1}$$
$$O^+ + O_2 \rightarrow O_2^+ + O, \gamma_4 = 9 \cdot 10^{-10} T^{-0.7}$$

A weight-averaged coefficient

$$\gamma = \frac{1}{M} (\gamma_3[N_2] + \gamma_4[O_2]).$$

is introduced in the last two reactions.

The processes considered in the calculations include the dissociative recombination of the molecular ions M^+ with a weight-averaged coefficient α, and the radiative recombination of the atomic ions O^+ and H^+, both with the coefficient, α, and the radiative recombination of the atomic ions O^+ and H^+, both with the coefficient, $\alpha = 10^{-12}$ cm³/sec.

The daytime recombination coefficient α is estimated in the E-layer maximum from the relationship $\alpha = q/n_c^2$. Its nighttime value in the E-layer maximum is considered to be temporally constant, and equal to the value just before sunset. Allowance is also made for the variation of α with the altitude, due to the change in the electron temperature given by an empirical formula.

Since the nighttime values of α and n_c in the E-layer maximum are known, the ion formation rate $q^m = \alpha n^2$ can be determined, while the variation of q with the altitude is assumed to be described by a simple Chapman layer. The calculation is performed under the following boundary conditions: at 100 km altitiude a photochemical equilibrium is assumed; at 1000 km altitude a diffusional equilibrium, i.e., the absence of plasma flux, is postulated.

The statistical part is the method of processing the ionospheric data provided by the global network of ionospheric sounding stations. The monthly median values of the diurnal variation of f_oF2, h_pF2 and f_oE obtained by 100–120 vertical sounding stations in 1962 and 1964 – years of low solar activity – are employed. Statistical processing of the initial data yields the global distribution of f_oF2, f_oE and h_pF2. These data are used to correct the theoretical calculations.

Profile correction. We shall see how to effect the correction of the calculated profile in the F2-region, since the profile $n_c(h)$ is not corrected in the E-region. The calculated values of n_c^m and h_m are compared with the experimental values, on the assumtpion that the reduced maximum altitude h_p is equal to the true altitude h_m. The deviations of the calculated maximum altitude from the experimental (Δh_m), as well as the coefficient a_m, the ratio between the experimental and the calculated electron concentrations in the layer maximum, are computed. Above the layer maximum the entire computed profile is shifted by Δh_m, while the concentration is changed by the factor a_m. In the E-region the profile is not corrected ($\alpha = 1$), and a linear interpolation of the coefficient, a, between the minimum (the trough between E- and F2-regions) and the maximum points of the F2-layer is carried out.

In addition to this method, a physical method of correction is also used. The calculated altitude h_m is adjusted to the experimental value with the aid of the vertical drift, and if the calculated value of the electron concentration differs from the experimental value, the coefficient γ is multiplied by a_m, and the calculation is repeated. This method is used in daytime. At night, in addition to the adjustment of the altitude with the aid of the drift, the upper boundary flux and the coefficient γ are changed. Two or three iterations are usually sufficient to obtain the desired accuracy – $\Delta h_m < 4$ km, and $\Delta n_c^m \leq 10\%$.

The *comparison with the experimental data* is conducted separately for the statistical and the theoretical parts of the model, using both the averaged ionospheric sounding data and the results of individual experiments performed on undisturbed days. An estimate was made of the accuracy of the model in representing the initial data at the various stations, and in reproducing the f_oF2-field for the subsequent solar activity cycle. For most stations the error does not exceed 5% by day or 10% at night. However, in individual regions the errors may be as large as 20% by day and up to 30% at night. A comparison of the data furnished by this model with those from the "Prediction of MUF" model indicate that both of them have about the same prediction accuracy for median critical frequencies f_oF2.

While the statistical part of the model is constructed from the data given by the global network of stations, and is inherently planetary, the theoretical part is limited by a belt of middle latitudes, since additional mechanisms must be considered in describing the polar and the equatorial zones. For this reason, a comparison between calculation and observations can only be made for mid-latitude electron concentration profiles. The authors report comparisons between the observed and the calculated profiles $n_c(h)$, obtained by the incoherent scatter method on individual dates at minimum solar activity.

The calculated profiles, corrected as described above with the aid of the observed n_c^m and h_m-values, agree well with the experimental profiles in many cases, especially so below the layer maximum. In order to estimate the accuracy of the model as a whole, the profile is calculated for a given point and a given moment of time, introducing the correction from the statistical part of the model. Since the correction is carried out with median values, comparison must also be made against the median values of electron concentration. The results of vertical sounding, carried out at the Irkutsk station in March 1962, were used for this purpose. Median values from the ionograms were compiled for the same times of day, and then used to calculate the profiles; these in turn were used in the comparison (Fig. 5.4). The principal differences in the profiles are due to discrepancies between the critical frequencies directly determined at the stations and those taken from the statistical part of the model.

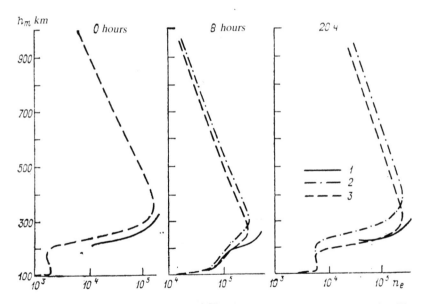

Fig. 5.4. Monthly median profiles, $n_c(h)$, for March 1962 at the Irkutsk station (1), uncorrected profiles
(2) and corrected profiles (3) /63/.
hours

Since the statistical part of the model makes it possible to determine the parameters f_oE, f_oF2 and h_pF2 for any point in space and at any moment in time and, in addition, the profile of electron concentration $N(h)$ may be calculated for mid-latitudes, the model may be employed in the computation of radiowave trajectories.

The Nisbet model /224/. This semiempirical model also consists of a statistical and a theoretical part, and a comparison between calculated values with those determined by observation.

The statistical part includes the data of the global distribution of the critical frequencies of the *F2*-layer in all existing vertical sounding stations during 1954–1958. The variation of the electron concentration $n_e^m F2$ with the level of the solar activity is a linear function of the index $F_{10.7}$.

In contrast to the Irkutsk model, the statistical part of the Nisbet model consists of data on the true latitudes of the *F2*-layer, obtained by processing ionograms from 8 medium- and low-altitude stations in the western part of the northern hemisphere from 1957–1962. Since the number of the data available is clearly insufficient to give a planetary respresentation of the h_m-values, sinusoidal approximations are employed to describe the annual and the latitudinal variations of h_m. It is assumed: that the principal component of the annual variations of h_m is seasonal, with a maximum in summer, and therefore the data are grouped separately for the winter (November, December, January) and the summer (May, June, July) months; and that the situations in the southern and northern hemispheres are identical. The diurnal variations of h_m are then represented as a Fourier series, and the summer and the winter coefficients may be determined for two solar activity levels $F_{10.7}$ (100 and 200). The h_m-values for intermediate $F_{10.7}$ levels may be found by linear interpolation.

The *theoretical part* is based on the thermospheric CIRA 1965 model /175/. The rates of formation of $q(O^+)$, $q(O_2^+)$, $q(N_2^+)$ ions are calculated from the Hinteregger spectrum and its variation with the solar activity according to /170/.

Three major ion-molecule reactions were considered:

$$O^+ + N_2 \rightarrow NO^+ + N \qquad \gamma = (2 \pm 1) \cdot 10^{-12} \text{ cm}^3/\text{sec};$$
$$O^+ + O_2 \rightarrow O_2^+ + O \qquad \gamma = (2 \pm 1) \cdot 10^{-11} \text{ cm}^3/\text{sec};$$
$$N_2^+ + O \rightarrow NO^+ + N \qquad \gamma = 2 \cdot 10^{-10} \text{ cm}^3/\text{sec},$$

and two dissociative recombination processes:

$$NO^+ e \rightarrow N + O \qquad \alpha = 1.44 \cdot 10^{-4}/T \text{ cm}^3/c;$$
$$O_2^+ + e \rightarrow O + O \qquad \alpha = 6.6 \cdot 10^{-5}/T \text{ cm}^3/\text{sec}.$$

The electron temperature in the layer maximum is computed from local equilibrium conditions between the rates of heating and cooling of the electron gas. It is assumed that T_e is independent of the altitude above the layer maximum. The profile of electron concentration is obtained by solving the stationary diffusion equation, while the altitude h_m and the concentration n_e^m are assigned empirical values, taken from the statistical part of the model. The contribution of the H^+ ions is evaluated by using experimental data for the electron concentration at 1000 km altitude and plotting a smooth profile for the altitude range h_m up to 1000 km. The electron concentration at 1000 km altitude is taken on the basis of empirical data for winter and summer at two $F_{10.7}$ solar activity levels – 100 and 200.

The calculated h_m and n_e^m-values are then compared to the experimental values and the rate constants for the first two ion-molecule reactions mentioned above and of the

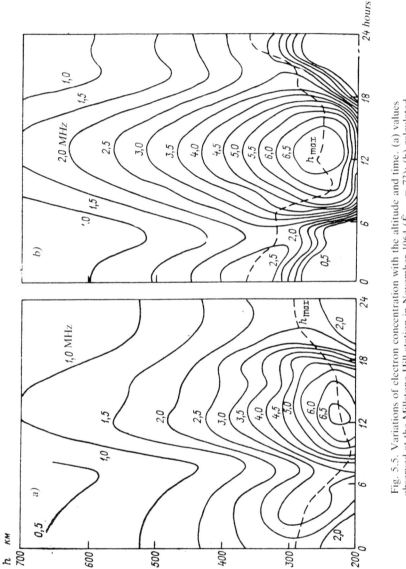

Fig. 5.5. Variations of electron concentration with the altitude and time. (a) values observed at the Millstone Hill station in November 1964 ($F_{10.7} = 73$); (b) calculated values.

diffusional flux value are selected (correction method). This method is used in daytime. The nighttime $n_e(h)$ profile is determined by empirical relationships depending on the atomic oxygen scale height and the values of the electron concentration at 1000 km altitude.

Comparison with the experiment is conducted by using both vertical sounding data at heights below the maximum layer of the F2-region, and the results of ionospheric observations made by the incoherent scatter method. In Fig. 5.5 a comparison of this type is shown for the Millstone Hill station in November 1964 ($\bar{F}_{10.7} = 73$). It is apparent that the diurnal variations of electron concentration in the region of the maximum are reproduced quite satisfactorily, but large discrepancies remain between the calculated and the observed layer altitudes. This difference is 50 km around noon, and may be as large as 70 km at midnight. The experimentally observed decrease of the ion concentration with altitude is also larger than the calculated value. The same is true for the month of July 1964, chosen for comparison purposes. The study also contains comparisons of the observation data with the calculated values on a number of specific days.

Let us consider the general advantages and drawbacks of the semiempirical models. As compared to statistical models of the "MUF Prediction" type the semiempirical models are better and more complete. They include statistically processed data on the global distribution of median values of critical frequencies; and they attempt to allow for the altitude of the layer maximum, and for the altitude distribution of electron concentration, i.e., for the $n_e(h)$ profile, required for solving problems of radiowave propagation. However, there are far fewer data available on the true altitudes h_m than on critical frequencies, making it impossible to give a global distribution of median h_m-values similar to the charts of f_oF2 for different local times, months and levels of solar activity. The Irkutsk model therefore suggests the use of data for the reduced altitudes, h_pF2, which are sufficiently numerous to make statistical treatment valid. However, the reduced altitude may be much higher (by up to 50 km) than the true altitude, so that a distorted altitude distribution of electron concentration may be obtained if the corrections are based on such data. It should also be noted that the theoretical parts of these models take processes typical of the mid-latitude F2-region into account, and therefore calculations of electron concentration profile and its subsequent correction can only be effected for mid-latitude regions.

A common feature of all these models is that they are based on monthly median values of f_oF2, which correlate well with the monthly average sunspot number, W. This is because W is an average index of the intensity of the short-wave solar flux I_o, as explained above. Consequently, these models can only describe median monthly average conditions in the F2-region. An attempt to describe isolated days, on which definite geophysical conditions prevailed, by such models may lead to large discrepancies as compared with the experimental values. This is to be expected, since W correlates with I_o only on the average, and not on individual days.

Finally we may note that the development of semiempirical models is an important step forward in the transition from purely statistical methods of describing the ionosphere to the deterministic methods based on rigorous analysis of the physical processes taking place in the thermosphere. With the aid of semiempirical models it is possible to proceed from the prediction of one parameter (critical frequency) to another (electron concentration profile). As the number of the available empirical data increases, the models may be appropriately updated.

5.3 DETERMINISTIC PREDICTION OF THE MID-LATITUDE IONOSPHERE

In contrast to the prediction methods discussed above, which involve the use of statistical and semiempirical models, based on ionospheric observation data, the deterministic approach describes the state of the ionosphere on the basis of physical processes taking place in the ionosphere. The results of the calculations are not corrected in accordance with the experimental data, and the principal test parameters are the indexes of solar and geomagnetic activity, which determine the state of the thermosphere and the flux of short-wave solar radiation. In principle, the neutral composition and the temperature of the upper atmosphere may also be computed from the corpuscular and short-wave solar radiation fluxes, but it is not yet possible to do this accurately enough, and therefore empirical models of the thermosphere, in which these relationships are postulated, are used in practical work.

In the computation method to be discussed below, the MSIS model is used as the thermospheric model. It yields the neutral composition and the temperature for magnetically quiet and weakly disturbed conditions ($A_p \le 10$), and also for different levels of solar activity, as indicated by the 10.7 cm solar radiation flux.

It should be pointed out that the relative variations in the short-wave solar flux, which is responsible for the ionization of the upper atmosphere, are also determined from the variations in the $F_{10.7}$ index (cf. sec. 3.1). Thus, the input parameters in the computation method considered here are the indexes $F_{10.7}$ and A_p, the values being routinely supplied to the prediction centers. These indexes determine the helio-geophysical situation on a given day, making it possible to calculate the state of the ionosphere on that particular day, but not its average state. Since the principal parameters which determine the state of the ionosphere are the flux of ionizing solar radiation and the neutral composition of the thermosphere, the day-to-day variations of these parameters may yield the ionosphere variations due to changed helio-geophysical conditions.

Thus, in order to be able to calculate the electron concentration in the $F2$-region under given conditions, the following indexes must be known: 1) the $\bar{F}_{10.7}$ index, averaged over three rotations of the Sun, the day in question being in the middle of this time interval; 2) the $F_{10.7}$ index for the preceding day; 3) the A_p index shifted by 6 hours with respect to the given time; and 4) the $F_{10.7}$ index for the day in question. The first three indexes determine the state of the thermosphere, while the fourth index gives the short-wave solar flux (cf. sec. 3.1). The input data also include the geographical latitude of the site and the serial number of the day in the year, which both determine the zenith angle of the Sun and the thermospheric parameters, as well as the magnetic dip and magnetic inclination of the site in question, required for the calculation of the coefficient of diffusion, the coefficient of thermal conduction, and the velocity of the vertical drift induced by the thermospheric winds. The next step is the solution of the initial set of equations, as described in sec. 2.2. This calculation yields the diurnal variations of the parameters of the neutral atmosphere – composition, temperature, thermospheric winds and, for certain thermospheric models, data on the ionized

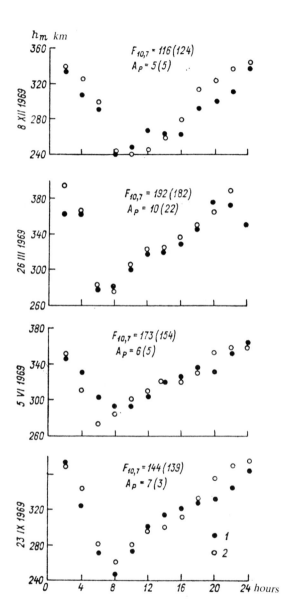

Fig. 5.6. Observed (1) and calculated (2) diurnal variations of the altitude h_m of the layer maximum in accordance with the data obtained at Millstone Hill station on four specific dates, for various values of the indexes $F_{10.7}$ and A_p, recorded on the same dates, respectively.

components as well, viz., electron concentration, ion composition, and plasma temperatures.

It is of interest to compare the results of the computations with the results of observations of the $F2$-region made on definite dates, under known helio-geophysical conditions. Figures 5.6 and 5.7 show the results of the calculations and observations of the diurnal variations of the altitude h_m and electron concentration n_c^m in the layer maximum on 4 days, in different seasons, and at different levels of geomagnetic and solar activity. It is apparent that the solar activity varied greatly – between $F_{10.7} = 116$ on 8th December 1969 and $F_{10.7} = 192$ on 26th March 1969 – all magnetically quiet days. The experimental data employed were those observed for the $F2$-region on these dates at the incoherent scatter station at Millstone Hill, and at ionospheric sounding stations in Moscow and Irkutsk. Data on the layer altitude h_m are from Millstone Hill only. For the results of comparison of the calculated values of h_m and n_c^m with the observations made on 32 dates see sec. 4.2.

It is important to note that the calculation not only gives a fairly accurate description of the diurnal variations of n_c, but also describes the variations in the altitude of the layer maximum. It may be seen from Fig. 5.6 (see also Fig. 4.14) that the differences between the calculated and the observed values of h_m in daytime do not exceed 20 km, but this difference may be larger at sunrise or sunset. Also, as was pointed out in sec. 4.2, there is a tendency for the calculated values of h_m to be higher during the evening and at nighttime. Even so, the maximum relative error of the calculated h_m-values at definite times never exceeds 15%. Comparison of the critical frequencies, f_o (sec. 4.2), showed that the computation accuracy in this case is 20–30%

Let us now consider the relative advantages and drawbacks of the deterministic calculation of the $F2$-region, as compared with the statistical and semiempirical methods. One of the main advantages of the deterministic approach is the utilization of the specific helio-geophysical data on the day in question. In practice, this may improve the quality of short-term prediction. Thus, if this method is adopted, the median value of f_oF2 taken over the past 10 days may be obtained once every 5 days. If, during the five days in question, the observed f_oF2-values have changed, this is considered as an ionospheric disturbance unrelated to the geomagnetic activity. It is possible to allow for a similar change in the level of solar activity, when giving a corrected median value.

Figure 5.8 shows such a case, which occurred during a rise in the solar activity in 1978. The median was taken over a 10-day period with an average solar activity level $\bar{F}_{10.7} = 160$, close to the value $\bar{F}_{10.7} = 169$, recorded on 5th December 1978. For this reason the observed critical frequencies are close to the median value, and the calculated values are close to those actually observed. During the daytime on 7th December 1978 the solar activity increased to $F_{10.7} = 186$, which would be expected to affect the rate of ion formation, due to an approximate 13% increase in the solar flux. Since the neutral composition and the temperature remained the same as on 5th December 1978 ($\bar{F}_{10.7} = 169$), the increase in the daytime electron concentration was proportional to the increase in the radiant flux, as may be seen in Fig. 5.8. It also follows from the figure that the calculation is carried out on the assumption that the above mechanism actually reproduces the observed increase of critical frequencies.

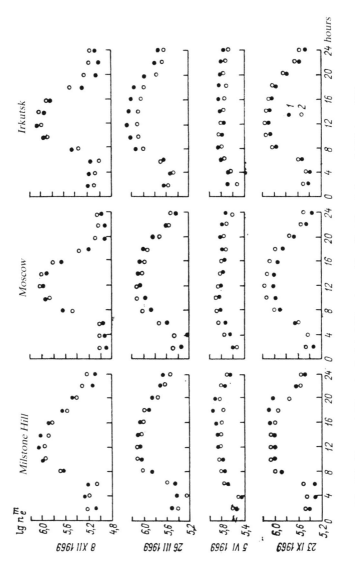

Fig. 5.7. Observed (1) and calculated (2) values of lg n_e^m at the stations of Millstone Hill, Moscow and Irkutsk on four different dates.

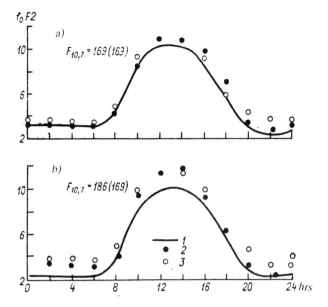

Fig. 5.8. Observed (1) and calculated (2) values of f_oF2
(MHz) in Moscow on the 5th December 1978 (a) and on
the 7th December 1978 (b). 1 – median; 2 – observations;
3 – calculation with allowance for the changed solar
activity level.

This method may be employed to calculate the monthly median distribution of the
electron concentration corresponding to magnetically quiet conditions for the monthly average value of $F_{10.7}$. Section 4.1 quotes examples of such calculations of the
annual variations of f_oF2 for the Northern and Southern hemispheres at different

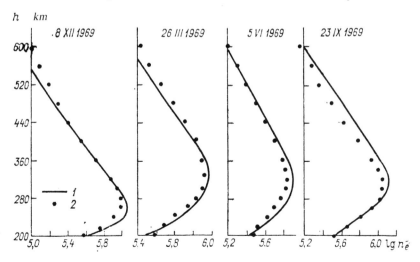

Fig. 5.9. Observed (1) and calculated (2) electron concentration profiles at 14 hrs at
the Millstone Hill station. The dates are the same as in Fig. 5.7.

solar activity levels as compared with the median values of f_0F2 in the "Prediction of MUF" model /73/.

An important feature of the deterministic method is that once the system of aeronomic parameters has been selected on the basis of reliable and comprehensive data on the $F2$-region, collected at one or more stations, the method may be used to calculate the $F2$-region at other points of the Earth, since the formation mechanism of the mid-latitude $F2$-layer is the same everywhere. A similar approach may be used for equatorial and auroral zones.

We shall illustrate this point by calculating the diurnal variations of the $F2$-layer at stations located in the Eastern and Western hemispheres. Since reliable data on h_m are lacking, comparison can only be carried out for the critical frequencies f_0F2, recorded at ionospheric sounding stations. We shall compare the calculated f_0F2-values with those observed on the same dates at Millstone Hill, Moscow and Irkutsk stations. It is apparent that the differences in the nighttime and daytime values of the critical frequency do not usually exceed 20% (Fig. 5.7).

We have seen that one of the advantages of the determinsitic approach may be used to calculate not only the parameters of the layer maximum, but also the electron concentration profile. Figure 5.9 shows the calculated and the observed $n_e^m(h)$ profiles at 14 hrs. There is full agreement between these two sets of data. However, the altitude distribution of electron concentration below about 200 km should be the subject of a separate investigation, aimed at clarifying the formation mechanism of the $F1$-zone, and the trough above the E-layer. These problems will not be discussed in this book.

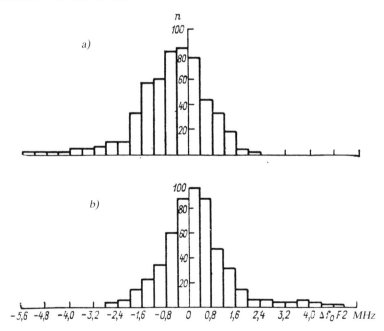

Fig. 5.10. Histograms of the differences Δf_0F2, obtained by the deterministic (a) and the statistical (b) methods.
/1 – MHz;/

The simplicity of calculation of the $F2$-region is one of the advantages of the method. In fact, the diurnal variations of the desired parameters can be obtained by a simple input of the values of a few well-known helio-geophysical indexes, as specified above. Short-term predictions are admittedly more complex, since a monthly average value of $\bar{F}_{10.7}$ is not available. For practical purposes, however, a sufficiently reliable value of $\bar{F}_{10.7}$ for the 1½ months that follow may be obtained by the method of inertial prediction, making allowance for the observed variation gradients of $F_{10.7}$

In practical work, the method can only be applied to the prediction of the state of the $F2$-region 1–2 days ahead, since we as yet unable to reliably predict the values of the input indexes, $F_{10.7}$ and A_p, for a longer term.

As an example, let us consider the results of prediction of f_oF2 at the Moscow station, one day ahead, as compared with the currently employed method based on the use of the "moving median". The processing was carried out on 89 magnetically quiet days ($A_p \leq 15$) between July and November 1979. In the deterministic calculation the predicted values of $F_{10.7}$ and A_p were used. The differences Δf_oF2 between the observed and the predicted values (534 differences in all) were determined for six points in time (0, 4, 8, 12, 16 and 20 hrs).

Figure 5.10 shows the histograms of the deviations Δf_oF2 (observed value – predicted value) for the f_oF2-values obtained by the deterministic and the statistical methods.

Table 5.2 shows the results of statistical processing of these data: mean deviation of Δf_oF2, mean square error, δf_oF2, and mean relative error, σ. It may be seen that, for the days selected, the three statistical indexes are similar for both methods, and that the mean relative error σ in the prediction of f_oF2 is 12–15%. It is also clear from Fig. 5.10 that the distribution of Δf_oF2 is close to normal in both cases.

TABLE 5.2

**Results of predictions of f_oF2 and h_m by the deterministic and
by the statistical methods.**

Parameter	Method	
	deterministic	statistical
Δf_oF2 MHz	−0.57	0.23
δf_oF2 MHz	1.17	1.27
σ $^\theta/_\theta$	15	12
Δh_m km	−10.1	−17.8
δh_m km	18.9	30.0
σ $^\theta/_\theta$	5.7	9

However, as already seen, the deterministic calculation yields several parameters which are not at present predicted, in particular the altitude of the $F2$ layer maximum. Since the h_m-values observed at the Moscow station are not available for the period in question, the comparison may be based, for example, on an epignosis of about 30 dates, for which observation data are available from the Millstone Hill station. The differences Δh_m (observed – calculated), were determined at four times of the day – 0, 6, 12 and 18 hrs – 132 differences altogether.

Figure 5.11 shows the histograms of the differences Δh_m obtained by the deterministic method and with the aid of model /4/, which may be used to predict h_m.

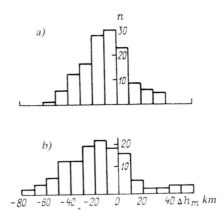

Fig. 5.11. Histograms of the differences, $\Delta h_c{}^m$ for values obtained by the deterministic method (1) and using the model /4/ (b).

It follows from Table 5.2 that the mean square error and the mean average error of h_m are lower by a factor of 1.5 lower when the determinstic method is employed. In addition, the mean deviation is smaller (10.1 as against 17.8 km) and the shape of the Δh_m histogram is also more satisfactory.

Thus, it is possible to predict the parameters of the $F2$-region by using the deterministic method. Comparison of the predicted with the observed values shows that predictions based on the deterministic approach are just as satisfactory as the statistical predictions, and even more so as far as the parameter h_m is concerned.

It should be noted, however, that a number of basic difficulties are involved. As mentioned above, short-wave solar radiation cannot as yet be predicted, and is now merely estimated from the $F_{10.7}$ index, which can be measured 2–3 days ahead at most. The same applies to the index, A_p, and the accuracy of prediction of this parameter, even 1–2 days ahead, is relatively low (about 30%).

There are also several other limitations, which are not directly connected with the proposed method, but are due to our imperfect knowledge of the upper atmosphere – and these will probably be eliminated in future. We have repeatedly stated that one of the key parameters is the model of the neutral atmosphere, which determines the composition and the temperature of the upper atmosphere and the system of thermospheric winds. Even though the empirical models of the thermosphere recently proposed represent a very considerable step forward, there are still difficulties in describing the variations of the neutral composition, concentrations of atomic and of molecular oxygen (cf. Chapter 3), and temperature and pressure gradients. The special features of certain regions – such as the geomagentic anomaly zone in Brazil, where the annual variations observed are not typical for the mid-latitude $F2$-region – have been totally ignored in thermospheric models /289/.

The description of the state of the atmosphere during disturbed periods – in particular ionospheric states during magnetic storms, a very timely problem, since such disturbances are often noted during the periods of medium-to-high and high solar activity – is a difficult problem indeed. There are also other difficulties which should be taken into consideration, such as the lack of sufficiently complete and reliable information about the electric fields in the $F2$-region, especially so during disturbed periods. These problems require further study.

Nevertheless, even now, the proposed deterministic method, used in conjunction with the statistical methods of predicting the atmosphere, can improve the quality of the prediction, and make it more complete. In the future, as the thermospheric models and the prediction of short-wave and corpuscular solar radiations are improved by means of routine observations of the Sun from satellites, deterministic prediction of the ionosphere will also become more accurate.

REFERENCES

1. Antonova, L.A. and Ivanov-Kholodnyi, G.S. Calculation of the relative amounts of ions at 140–200 km altitudes under specific helio-geophysical conditions.– Geomagn. i aeronomiya, Vol. 18: pp. 837–843 (1978).

2. Antonova, L.A. and Katyushina, V.V. A possible mechanism of the genesis of semiannual density variations in the upper atmosphere.–Geomagn. i aeronomiya, 16(2): pp. 311–315 (1976).

3. Antonova, L.A. and Katyushina, V.V. Seasonal latitude variations of the turbulence and of the neutral composition of the upper atmosphere.–Geomagn. i aeronomiya, 20(1): pp. 67–71 (1980).

4. Anufrieva, T.A. and Shapiro, B.S. Book: "Geometricheskie parametry sloya $F2$ ionosfery" (Geometrical Parameters of the Ionospheric $F2$-Layer).– Moscow, "Nauka" Publishing House, 1976.

5. Besprozvannaya, A.S. The nature of the seasonal anomaly of the noontime ionization of the $F2$-layer.–Geomagn. i aeronomiya, 10(6): pp. 986–992 (1970).

6. Boenkova, N.M. and Mikhailov, A.V. The effect of December anomaly in the $F2$-region during the solar activity cycle.–Geomagn i aeronomiys, 20(3): pp. 445–448 (1980).

7. Vasil'eva, T.N. et al. Estimate of the accuracy of the long-term prediction of median monthly values of f_oF2.–Geomagn. i aeronomiya, 16(4): pp. 633–637 (1976).

8. kVlasov, M.N., Chernyshev, V.N. and Kolesnik, V.G. Distribution of molecular nitrogen over vibrational levels in the upper atmosphere.–Geomagn. i aeronomiya, 18(4): pp. 645–651 (1978).

9. Vsekhsyatskaya, I.S., Sergeenko, N.P. and Yudovich, L.A. Possibilities of statical modelling of critical $F2$-layer frequencies.–In book: "Ionosfernye vozmushcheniya i metody ikh prognoza", Moscow, "Nauka" Publishing House, pp. 3–9.

10. Gershengorn, G.E. Certain numerical methods for solving the equation of electron diffusion in the ionosphere.–Issled. po geomagn., aeronomii i fizike Solntsa, No. 21: pp. 283–290 (1972).

11. Danilov, A.D. Book: "Khimiya ionosfery" (The Chemistry of the Ionosphere)–Leningrad, "Gidrometeoizdat" Publishing House, 1967.

12. Danilov, A.D. and Ivanov-Kholodnyi, G.S. Experimental data on the intensities of energy sources in the ionosphere.–Geomagn. i aeronomiya, 3(5): pp. 850–857 (1963).

13. Johnson, F.S. Variations in the composition of upper atomsphere.–In book: "Electron Concentration in the Ionosphere and in the Exosphere" [Russian translation].

14. Zevakina, R.A. et al. Methods of short-term prediction of magnetic activity and the state of the ionosphere.–Instruktsiya IZMIRAN/R, Moscow, 1975.

15. Zevakina, R.A., Namgaladze, A.A. and Smertin, V.M. An interpretation of the positive disturbances of $F2$-region.–Geomagn. i aeronomiya, 18(6): pp. 1040–1044 (1978).

16. Ivanov-Kholodnyi, G.S. Short-wave solar radiation and certain problems in physics of the ionosphere.–Ph.D. Dissertation, Moscow, 1966.

17. Ivanov-Kholodnyi, G.S. Composition and structure of the upper atmosphere. In book: "Itogi nauki i tekhniki, ser. geomagn. i vysokie sloi atmosfery."–Moscow, VINITI, 1972.

18. Ivanov-Kholodnyi, G.S. Semiannual variations in aeronomy and geomagnetism.–Geomagn. i aeronomiya, 13(6): pp. 969–990 (1973).

19. Ivanov-Kholodnyi, G.S. Prediction of relative changes in the intensity of short-wave solar radiation during the eleven-year cycle.–Geomagn. i aeronomiya, 14(2): pp. 188–191 (1974).

20. Ivanov-Kholodnyi, G.S. Composition and structure of the upper atmosphere. In book: "Itogi nauki i tekhniki, ser. geomagn. i vysokie sloi atmosfery".–Moscow, VINITI, Vol. 3: pp. 7–61 (1976).

21. Ivanov-Kholodnyi, G.S. and Antonova, L.A. Variations of [NO^+] / [O_2^+] ratio in daytime at 130–200 km altitude.–Geomagn. i aeronomiya, 15(3): pp. 477–482 (1975).

22. Ivanov-Kholodnyi, G.S. and Velichanskii, B.N. A review of the data on the short-wave solar radiation.–Issledovaniya po geomagnetizmu, aeronomii i fizike Solntsa, No. 26: pp. 14–25 (1972).

23. Ivanov-Kholodnyi, G.S. and Katyushina, V.V. Estimates of the altitude and the concentration of the atomic oxygen layer in the upper atmosphere.–Geomagn. i aeronomiya, 11(5): pp. 819–1824 (1971).

24. Ivanov-Kholodnyi, G.S. and Katyushina, V.V. Semiannual variations of the atmospheric composition at 130–200 km altitudes.–Geomagn. i aeronomiya, 14(4): pp. 674–679 (1974).

25. Ivanov-Kholodnyi, G.S. and Firsov, V.V. The spectrum of short-wave solar radiation at various activity levels.–Geomagn. i aeronomiya, 14(3): pp. 393–398 (1974).

26. Ivanov-Kholodnyi, G.S. et al. Effect of seasonal variations of neutral atmosphere on the ionization of the E-region of the ionosphere.–Geomagn. i aeronomiya, 17(5): pp. 839–846 (1977).

27. Ivanov-Kholodnyi, G.S., Leshchenko, L.N. and Odintsova, I.N. The ratio between the X-ray and the UV radiation of the solar flares in the ionization of the E-region of the ionosphere.–Geomagn. i aeronomiya, 16(2): pp. 246–250 (1976).

28. Ivanov-Kholodnyi, G.S. and Mikhailov, A.V. Effect of the variable ionospheric-protonospheric plasma flux on the nighttime F-layer of the ionosphere.– Geomagn. i aeronomiya, 131(1): pp. 47–51 (1973).

29. Ivanov-Kholodnyi, G.S. and Mikhailov, A.V. A model of the seasonal and semiannual variations of the $F2$-region of the ionosphere.–Geomagn. i aeronomiya, 15(4): pp. 739–741 (1975).

30. Ivanov-Kholodnyi, G.S. and Mikhailov, A.V. Relationship between the parameters of the $F2$-region of the ionosphere and the parameters of the neutral atmosphere at a given altitude.–Geomagn. i aeronomiya, 16(5): pp. 799–802 (1976).

31. Ivanov-Kholodnyi, G.S. and Mikhailov, A.V. Selection of a system of parameters for modelling the daytime $F2$-region.–Geomagn. i aeronomiya, 17(1): pp. 30–34 (1977).

32. Ivanov-Kholodnyi, G.S. and Nikol'skii, G.M. Identification of the solar radiation lines in the short-wave spectral range – $\lambda < 1100$ A.–Geomagn. i aeronomiya, 2(3): pp. 425–442 (1962).

33. Ivanov-Kholodnyi, G.S. and Nikol'skii, G.M. Book: "Solntse i ionosfera" (The Sun and the Ionosphere).–Moscow, "Nauka" Publishing House, 1969.

34. Ivanov-Kholodnyi, G.S. and Nusinov, A.A. The relative contributions of UV and X-ray radiations to the ionization of the E-region.–Geomagn. i aeronomiya, 16(1): pp. 76–79 (1976).

35. Ivanov-Kholodnyi, G.S. and Nusinov, A.A. The formation and dynamics of the mid-latitude daytime E-layer of the ionosphere.–Turdy IPG, No. 37 (1979).

36. Ivel'skaya, M.K., Katyushina, V.V. and Klimov, N.N. A possible interconnection between semiannual and annual variations of the concentration of oxygen in the upper atmosphere.–Geomagn. i aeronomiya, 18(1): pp. 91–95 (1978).

37. Kazachevskaya, T.V. The seasonal anomaly of the $F2$-layer and variations in atmospheric composition.–Geomagn. i aeronomiya, 7(6): pp. 1096–1098 (1967).

38. Katyushina, V.V. and Ivanov-Kholodnyi, G.S. Semiannual variations in the concentration of oxygen.–Geomagn. i aeronomiya, 11(5): pp. 919–920 (1971).

39. King–Hille, D. Artificial satellites and scientific research.– [Russian translation], 1963.

40. Kolesnik, A.G. Semiannual variations of the neutral composition of the base of Earth's thermosphere.–Geomagn. i aeronomiya, 15(2): pp. 286–290 (1975).

41. Kolesnik, A.G. and Chernyshev, V.E. A non-stationary self-consistent model of the mid-latitude F-region of the ionosphere.–IV Inter–Departmental Seminar on the Modelling of the Ionosphere.–Tomsk, Tomsk University Publishing House, 1978.

42. Book: "Kosmicheskaya geofizika" (Cosmic Geophysics).–Moscow, "Mir" Publishing House, 1976.

43. Krinberg, I.A. Book: "Kinetika elektronov v ionosfere i plazmosfere Zemli" (The Kinetics of Electrons in the Ionosphere and Plasmosphere of the Earth).–Moscow, "Nauka" Publishing House, 1978.

44. Krinberg, I.A., Kuz'min, V.A. and Gershengorn, G.I. An ionospheric model allowing for plasma motion along geomagnetic field force lines.–Geomagn. i aeronomiya, 14(2): pp. 224–230 (1974).

45. Krinberg, I.A. Special features in the estimation of the interaction with the plasmosphere in theoretical models of the ionosphere. IV Inter-Departmental Seminar on the Modelling of the Ionosphere.–Tomsk, Tomsk University Publishing House, 1978.

46. Lavrova, E.V. Numerical method of ionospheric prediction. In book: "Ionosfernye vozmushcheniya i metody ikh prognoza", Moscow, "Nauka" Publishing House, 1977.

47. Leshchinskaya, T. Yu. Review of Ionospheric Models Employed in Solving Wave Propagation Problems. Propagation of decametric waves.–Moscow, IZMIR-AN, 1978.

48. L'vova, A.A., Polyakov, V.M. and Rybin, V.V. Special features of nonstationary solutions of the diffusion equation in a gravitational field.–Issl. po geomagn., aeronom. i fizike Solntsa, No. 18: pp. 3–24 (1978).

49. Lyakhova, L.N. and Kostina, L.I. Quantitative ionospheric prediction.–Geomagn. i aeronomiya, 13(1): pp. 59–63 (1973).

50. McIven, M. and Phillips, L. The Chemistry of the Atmosphere. [Russian translation], 1978.

51. Mkhailov, A.V. A check on the adequacy of the calculation system of a real mid-latitude noontime $F2$-region of the ionosphere.–Geomagn. i aeronomiya, 18(6): pp. 1029–1032 (1978).

52. Mikhailov, A.V. and N.M. Boenkova. Asymmetry of the annual variations in the $F2$-regions of the Northern and Southern hemispheres.–Geomagn. i aeronomiya, 20(2): 251–254 (1980).

53. Mikhailov, A.V. and Ostrovskii, G.E. Calculation of the nighttime, magnetically quiet $F2$-region of mid-latitude ionosphere.–Geomagn. i aeronomiya, 18(2): pp. 224–228 (1978).

54. Mikhailov, A.V. and Ostrovskii, G.E. Calculation of nighttime magnetically quiet mid-latitude F2-region of the ionosphere.–Geomagn. i aeronomiya, 18(2): pp. 224–228 (1978).

55. Mikhailov, A.V. and Ostrovskii, G.I. The effect of winter growth of electron concentration in the nighttime F2-region and its possible interpretation.– Geomagn. i aeronomiya, 20(1): pp. 29–32 (1980).

56. Mikhailov, A.V. and Serebryakov, B.E. A model of latitudinal variations of the annual evolution of n_e^m in the F2-region. Variation with the solar activity level.–Geomagn. i aeronomiya, 17(3): pp. 521–523 (1977).

57. Mikhailov, A.V. and Serebryakov, B.E. Calculation of the diurnal variations of the parameters of the mid-latitude, magnetically quiet ionospheric F2-layer maximum.–Geomagn. i aeronomiya, 19(6): pp. 1001–1007 (1979).

58. Nikitin, M.A. and Serebryakov, B.E. Variation of plasma flux at mid-latitudes at high solar activity levels.–Geomagn. i aeronomiya, 18(3): pp. 526–528 (1978).

59. Nikole, M. Book: "Svoistva i stroenie verkhnei atmosfery" (Properties and Structure of the Upper Atmosphere).–Moscow, "Fizmatgiz" Publishing House, 1963.

60. Nikol'skii, G.N. The energy of the short-wave solar radiation in the λ < 1100 A spectral range.–Geomagn. i aeronomiya, 3(5): pp. 793–809 (1963).

61. Ostrovskii, G.I. The relationship between the altitude of the nighttime F2-region and the aeronomic parameters.–Geomagn. i aeronomiya, 17(5): pp. 939–940 (1977).

62. Polyakov, V.M. and Rybin, V.V. Dynamic problems in ionospheric F-region. The Sturm–Liouville problem.–Geomagn. i aeronomiya, 15(5): pp. 806–812 (1975).

63. Polyakov, V.M. et al. A semi-empirical ionospheric model. In book: "Materialy mirovogo tsentra dannykh B.".–Moscow, "Gidrometeoizdat" Publishing House, 1978.

64. Ratcliffe, J.A. and Weeks, K. Book: "Ionosphere. Physics of the Upper Atmosphere".–[Russian translation].

65. Rishbet, G. and Harriot, O.K. Book: "Introduction to the Physics of the Ionosphere".–[Russian translation].

66. Sazhin, V.I. The utilization of the hybrid model of the ionosphere in the computer program for the calculation of the radiowave propagation parameters.– Issled. po geomagn., aeronom. i fizike Solntsa, No. 41: pp. 117–120 (1977).

67. Samarskii, A.A. Book: "Vvedeni v teoriyu raznostnykh skhem" (Introduction to the Theory of Difference Routines).–Moscow, "Nauka" Publishing House, 1973.

68. Instructions for the Utilization of Monthly Predictions of MUF.–Leningrad, "Gidrometeoizdat" Publishing House, 1976.

69. Fatkullin, M.N. Theoretical models of the seasonal variations of electron concentration in the mid-latitude $F2$-region.–Geomagn. i aeronomiya, 15(2): pp. 246–250 (1975).

70. Book: "Fizika verkhnei atmosfery" (Physics of the Upper Atmosphere).–Moscow, "Fizmatgiz" Publishing House, 1963.

71. Chapman, S. and Cowling, T. Book: "Mathematical Theory of In-homogeneous Gases".–[Russian translation].

72. Chernyshev, V.I. The periodicity of the variations of the UV solar radiation.–Geomagn. i aeronomiya, 18(5): pp. 798–803 (1978).

73. Chernyshev, O.V., and Vasil'eva, T.N. Book: "Prognoz maximal'no primenimykh chastot" (Prediction of MUFs).–Moscow, "Nauka" Publishing House, 1975.

74. Chernyshev, O.V. and Shapiro, B.S. Analytical description of the charts of the geometric parameters of the $F2$-layer of the ionosphere by means of spherical functions.–Geomagn. i aeronomiya, 17(6): pp. 1111–1112 (1977).

75. Abur–Robb, M.F.K. Combined world–wide neutral air wind and electro-dynamic drift effects on the $F2$-layer.–Planet. Space Sci., 17(6): pp. 1269–1279 (1969).

76. Albritton, D.L., Dotan, I., Lindinger, W., and McFarland, M. Effects of ion speed distributions in flow-drift tube studies of ion-neutral reactions.–J.Chem. Phys., 66(2): pp. 410–421 (1977).

77. Alcayde, D., Bauer, P., and Fontanari, J. Long-term variations of thermospheric temperature and composition.–J. Geophys. Res., 79(4): pp. 629–637 (1974).

78. Amayenc, P., Alcayde, D., and Kockarts, G. Solar extreme ultraviolet heating and dynamical processes in the mid-latitude thermosphere.–J. Geophys. Res., 80(19): pp. 2887–2891 (1975).

79. Amayenc, P., and Vasseur, G. Neutral winds deduced from incoherent scatter observations and their theoretical interpretation.–J. Atmos. Terr. Phys., 34(3): pp. 351–364 (1972).

80. Antoniadis, D.A. Thermospheric winds and exospheric temperatures from incoherent scatter radar measurements in four seasons.–J. Atmos. Terr. Phys., 38(2): pp. 187–195 (1976).

81. Axford, W.I. The polar wind and the terrestrial helium budget.–J. Geophys. Res., 73(21): pp. 6855–6859 (1968).

82. Baily, G.J., Moffett, R.J., and Murphy, J.A. A theoretical study of night-time field-aligned O^+ and H^+ fluxes in a mid-latitude magnetic field tube at equinox under sunsport maximum conditions.–J. Atmos. Terr. Phys., 39(1): pp. 105–110 (1977).

83. Banks, P.M. Electron thermal conductivity in the ionosphere.–Earth Planet. Sci. Lett., 1(1): pp. 151–154 (1966).

84. Banks, P.M. Ion temperature in the upper atmosphere.–J. Geophys. Res., 72(13): pp. 3365–3385 (1967).

85. Banks, P.M. The temperature coupling of ions in the ionosphere.–Planet. Space Sci., 15(1): pp. 77–93 (1967).

86. Banks, P.M. The thermal structure of the ionosphere.–Proc. IEEE, 57(3): pp. 258–281 (1969).

87. Banks, P.M. Magnetospheric processes and the behavior the neutral atmosphere.–Space Research XII, Vol. 2: pp. 1051–1067 (1972).

88. Banks, P.M., and Holzer, T.E. The polar wind.–J. Geophys. Res., 73(21): pp. 6846–6854 (1968).

89. Banks, P.M. and Holzer, T.E. Features of plasma transport in the upper atmosphere.–J. Geophys. Res., 74(26): pp. 6304–6316 (1969).

90. Banks, P.M., and Holzer, T.E. High-latitude plasma transport: the polar wind.–J. Geophys. Res. 74(26): pp. 6317–6332 (1969).

91. Banks, P.M. and Kockarts, G. Aeronomy, Part B.–Academic Press, New York–London, p. 355, 1973.

92. Banks, P.M., Nagy, A.F., and Axford, W.J. Dynamical behavior of thermal protons in the mid-latitude ionosphere and magnetosphere.– Planet. Space Sci. 19(9): pp. 1053–1067 (1971).

93. Barker, J.R. Radio astronomical measurements of ionospheric electron content.–J. Atmos. Terr. Phys., 34(11): pp. 1923–1933 (1972).

94. Barlier, F., Banner, P., Zaeck, C., Thuillier, G., and Kockarts, G. North–south asymmetries in the thermosphere during the last maximum of the solar cycle.–J. Geophys. Res., 79(34): pp. 5273–5285 (1974).

95. Barlier, F., Berger, C., Falin, J.L., Kockarts, G., and Thuillier G. A thermospheric model based on satellite drag data.–Ann. Geophys. 34(1): pp. 9–24 (1978).

96. Barrington, R.E. Ionospheric ion composition deduced from VLF observations.–Proc. IEEE, 57(6): pp. 1036–1041 (1969).

97. Berkner, L.V., Wells, H.W., and Seaton, S.L. Characteristics of the upper region of the ionosphere.–Terr. Magnet. Atmos. Elect., Vol. 41, June 6, pp. 173–184 (1936).

98. Bertin, F., and Lepine, J.P. Latitudinal variation of total electron content in the winter at midlatitudes.–Radio Sci., 5(6): pp. 899–906 (1970).

99. Bilitza, D. Models for the relationship between electron density and temperature in the upper ionosphere.–J. Atmos. Terr. Phys., 37(9): pp. 1219–1222 (1975).

100. Biondi, M.A. Atmospheric electron-ion and ion-ion recombination processes.–Canad. J. Chem., 47(5): pp. 1711–1719 (1969).

101. Blamont, J.E., Luton, J.M., and Nisbet, J.S. Global temperature distributions from OGO–6 6300 A airglow measurements.–Radio Sci., 9(2): pp. 247–251 (1974).

102. Blanc, M., Amagenc, P., Bauer, P., and Taieb, C. Electric field induced drifts from the French incoherent scatter facility.–J. Geophys. Res., 82(1): pp. 87–97 (1977).

103. Blum, P.W., and Harris, I. Full non-linear treatment of the global thermospheric wind system. I. Mathematical method and analysis of forces.–J. Atmos. Terr. Phys. 37(2): pp. 193–212 (1975).

104. Bowhill, S.A. The formation of the daytime peak of the ionospheric $F2$-layer.–J. Atmos. Terr. Phys., Vol. 24: pp. 503–520 (1962).

105. Brinton, H.C. Implications for ionospheric chemistry and dynamics of a direct measurement of ion composition in the F-region.–J. Geophys. Res. 74(11): pp. 2941–2951 (1969).

106. Brinton, H.C., Grebowsky, J.M., and Mayr, H.G. Altitude variation of ion composition in the midlatitude trough region: evidence for upward plasma flow.–J. Geophys. Res., 76(16): pp. 3738–3745 (1971).

107. Carpenter, L.S., and Kirchhoff, V.W.J.H. Comparison of high-latitude and midlatitude ionospheric electric fields.–J. Geophys. Res., 80(13): pp. 1810–1814 (1975).

108. Carpenter, D.L. and Park, C.G. On what ionospheric workers should known about the plasmapause-plasmosphere.–Rev. Geophys. Space Phys., 11(1): pp. 133–154 (1973).

109. Challinor, R.A. Neutral-air winds in the ionospheric F-region for an asymmetric global pressure system.–Planet. Space Sci., 17(6): pp. 1097–1106 (1969).

110. Challinor, R.A., and Eccles, D. Longitudinal variations of the mid-latitude ionosphere produced by neutral-air winds. I. Neutral-air winds and ionospheric drifts in the northern and southern hemispheres.–J. Atmos. Terr. Phys., 33(3): pp. 363–369 (1971).

111. Champion, K.S.W. Dynamics and structure of the quiet thermosphere.–J. Atmos. Terr. Phys. 37(6/7): pp. 915–926 (1975).

112. Chandra, S., and Spencer, N.W. Exospheric temperatures inferred from the Aeros A neutral composition measurement.–J. Geophys. Res., 80(25): pp. 3615–3621 (1975).

113. Chandra, S., and Stubbe, P. On explaining the F-region seasonal anomaly in terms of composition changes in the lower atmosphere.–Planet. Space Sci., 19(8): pp. 1014–1016 (1971).

114. Chen, A., Johnsen, R., and Biondi, M.A. Measurements of the O^+ + N_2 and O^+ + O_2 reaction.–J. Chem. Phys., 69(6): pp. 2688–2691 (1978).

115. Cho, H.R., and Yeh, K.C. Neutral winds and the behavior of the ionospheric $F2$-region.–Radio Sci., 5(6): pp. 881–894 (1970).

116. Clark, D.H., Raitt, W.J., and Willmore, A.P. The global morphology of electron temperature in the topside ionosphere as measured by an a.c. Langmuir probe.–J. Atmos. Terr. Phys., 34(11): pp. 1865–1880 (1972).

117. Cole, K.D. Electrodynamic heating and movement of the thermosphere.–Planet. Space Sci., 19(1): pp. 59–75 (1971).

118. Cook, G.R., Whitson, M.E., and McNeal, R.J. Temperture dependence of the quenching of vibrationally excited N_2 by O.– Trans. AGU, 54(4): p. 403 (1973).

119. Cox, L.P., and Evans, J.V. Seasonal variation of the O/N_2 ratio in the $F1$-region.–J. Geophys. Res., 75(31): pp.6271–6286 (1970).

120. Dalgarno, A. The effect of oxygen cooling on ionospheric electron temperatures.–Planet. Space Sci., 16(11): pp. 1371–1380 (1968).

121. DaRosa, A.V., and Smith, F.L. Behavior of the nighttime ionosphere.–J. Geophys. Res., 72(7): pp. 1829–1836 (1967).

122. Delaboudiniere, J.P., Donnelly, R.F., Hinteregger, H.E., Schmidtke, G., and Simon, P.C. Intercomparison/compilation of relevant solar flux data related to aeronomy. First report to the Working Group IV of COSPAR, May 1976.

123. Delaboudiniere, J.P., Donnelly, R.F., Hinteregger, H.E., Schmidtke, G., and Simon, P.C. Intercomparison/compilation of relevant solar flux data related to aeronomy. Report to the Working Group IV of COSPAR, June 1977.

124. Donahue, T.M., Guenther, B., and Thomas, R.Z. Distribution of atomic oxygen in the upper atmosphere deduced from OGO–6 airglow observation.–J. Geophys. Res., 78(28): pp. 6662–6689 (1973).

125. Donahue, T.M., Guenther, B., and Thomas, R.Z. Spatial and temporal behavior of atomic oxygen determined by OGO–6 airglow observations.–J. Geophys. Res., 79(13): pp. 1959–1964 (1974).

126. Dungey, J.W. The effect of ambipolar diffusion in the nighttime F-layer.–J. Atmos. Terr. Phys., 9(2/3): pp. 90–102 (1956).

127. Dunkin, D.B., Fehsenfeld, F.C., Schmeltekopf, A.L., and Ferguson, E.E. Ion-molecule reaction studies from 300–600 K in a temperature-controlled flowing afterglow system.–J. Chem. Phys., 49(3): pp. 1365–1371 (1968).

128. Eccles, D. The semi-annual variation in the height of the $F2$-layer peak.–J. Atmos. Terr. Phys., 33(10): pp. 1641–1646 (1971).

129. Eccles, D. Enhancements of the electron concentration in the $F2$-layer at magnetic noon.–J. Atmos. Terr. Phys., 35(7): pp. 1309–1315 (1973).

130. Eccles, D. and Burge, J.D. The behaviour of the upper ionosphere over north America at sunset.–J. Atmos. Terr. Phys., 35(11): pp. 1927–1934 (1973).

131. Eccles, D., King, J.W., and Kohl, H. Further investigations of the effects of neutral-air winds on the ionospheric F-layer.–J. Atmos. Terr. Phys., 33(9): pp. 1371–1381 (1971).

132. Evans, J.V. On the behavior of f_oF2 during solar eclipses.–J. Geophys. Res., 70(3): pp. 733–738 (1965).

133. Evans, J.V. Ionospheric backscatter observations at Millstone Hill.–Planet. Space Sci., 13(11): pp. 1031–1074 (1965).

134. Evans, J.V. Cause of the midlatitude winter night increase in f_oF2.–J. Geophys. Res., 70(17): pp. 4331–4345 (1965).

135. Evans, J.V. Cause of the mid-latitude evening increase in f_oF2.–J. Geophys. Res., 70(5): pp. 1175–1185 (1965).

136. Evans, J.V. Midlatitude F-region densities and temperatures at sunspot minimum.–Planet. Space Sci., 15(9): pp. 1387–1405 (1967).

137. Evans, J.V. Millstone–Hill Thomson scatter results for 1965.– Planet. Space Sci., 18(8): pp. 1225–1253 (1970).

138. Evans, J.V. F-region heating observed during the main phase of magnetic storms.–J. Geophys. Res., 75(25): pp. 4815–5823 (1970).

139. Evans, J.V. The June 1965 magnetic storm: Millstone–Hill observations.–J. Atmos. Terr. Phys., 32(10): pp. 1629–1640 (1970).

140. Evans, J.V. Observations of F-region vertical velocities at Millstone–Hill, evidence for drifts due to expansion, contraction and winds.–Radio Sci., 6(6): pp. 609–626 (1971).

141. Evans, J.V. Millstone–Hill Thomson scatter results for 1967. Lincoln Lab., M.I.I., Lexington, Mass. Tech. Rep., N. 482, 1971.

142. Evans, J.V. Ionospheric movements measured by incoherent scatter: A review.–J. Atmos. Terr. Phys., 34(2): pp. 175–209 (1972).

143. Evans, J.V. Millstone–Hill Thomson scatter results for 1966 and 1967.– Planet. Space Sci., 21(5): pp. 763–792 (1973).

144. Evans, J.V. Seasonal and sunspot cycle variations of F-region electron temperatures and protonospheric heat fluxes.–J. Geophys. Res., 78(13): pp. 2344–2349 (1973).

145. Evans, J.V. Millstone–Hill Thomson scatter results for 1968. Lincoln Lab., M.I.I., Lexington, Mass. Tech. Rep., N. 499 (1973).

146. Evans, J.V. Millstone–Hill Thomson scatter results for 1969. Lincoln Lab., M.I.I., Lexington, Mass. Tech. Rep., N. 513, 1974.

147. Evans, J.V. A study of $F2$-region night-time vertical ionization fluxes at Millstone–Hill.–Planet. Space Sci., 23(12): pp. 1611–1623 (1975).

148. Evans, J.V. Millstone–Hill Thomson scatter results for 1970. Lincoln Lab., M.I.I., Lexington, Mass. Tech. Rep., N. 522, 1976.

149. Evans, J.V., and Holt, J. Observations of F-region vertical velocities at Millstone–Hill. Determination of altitude distribution of H^+.–Radio Sci., 6(10): pp. 855–861 (1971).

150. Eyfrid, R.W. The effect of the magnetic declination on the $F2$-layer.–J. Geophys. Res., 68(9): pp. 2529–2530 (1963).

151. Fedder, J.A., and Banks, P.M. Convection electric fields and polar thermospheric winds.–J. Geophys. Res., 77(13): pp. 2328–2340 (1972).

152. Fehsenfeld, F.C., Dunkin, D.B., and Ferguson, E.E. Rate constants for the reaction of CO_2^+ with O, O_2 and NO; N_2^+ with O and NO and O_2^+ with NO.–Planet. Space Sci., 18(8): pp. 1267–1269 (1970).

153. Fehsenfeld, F.C., and Ferguson, E.E. Recent laboratory measurements of D- and E-region ion-neutral reactions.–Radio Sci., 7(1): pp. 113–15 (1972).

154. Ferguson, E.E. Ionospheric ion–molecule reaction rates.–Rev. Geophys. Space Phys., 5(3): pp. 305–312 (1967).

155. Ferguson, E.E. Laboratory measurements of F-region reaction rates.–Ann. Geophys., 25(3): pp. 819–823 (1969).

156. Ferguson, E.E. Laboratory measurements of ionospheric ion-molecule reaction rates.–Rev. Geophys. Space Phys., 12(4): pp. 703–713 (1974).

157. Ferguson, E.E., Bohme, D.K., Fehsenfeld, F.C., and Dunkin, D.B. Temperature dependence of slow ion-atom interchange reactions.–J. Chem. Phys., 50(11): pp. 5039–5040 (1969).

158. Fite, W.L. Positive ion reactions.–Can. J. Chem., 47(10): pp. 1797–1807 (1969).

159. Forbes, J.M. Wind estimates near 150 km from the variation in inclination of low perigee satellite orbits.–Planet. Space Sci., 23(4): pp. 726–731 (1975).

160. Frank, L.A., Ackerson, K.L. and Yeager, D.M. Observations of atomic oxygen (O^+) in the earth's magnetotail.–J. Geophys. Res., 82(1): pp. 129–134 (1977).

161. Geisler, J.E. Atmospheric winds in the middle latitude $F2$-region.–J. Atmos. Terr. Phys., 28(8): pp. 703–720 (1966).

162. Geisler, J.E. A numerical study of the wind system in the middle thermosphere.–J. Atmos. Terr. Phys., 29(12): pp. 1469–1482 (1967).

163. G e i s l e r, J.E. On the limiting daytime flux of ionization into the protonosphere.–J. Geophys. Res., 72(1): pp. 81–85 (1967).

164. G e i s l e r, J.E., and B o w h i l l, S.A. Ionospheric temperatures at sunspot minimum.–J. Atmos. Terr. Phys., 27(4): pp. 457–474 (1965).

165. G e i s l e r, J.E., and B o w h i l l, S.A. An investigation of ionosphere-protonosphere coupling.–Aeronomy Rep., N. 5: p. 250 (1965).

166. G l i d d o n, J.E.C., and K e n d a l l, P.C. A mathematical model of the $F2$-region.–J. Atmos. Terr. Phys., 24(12): pp. 1073–1100 (1962).

167. G o n z a l e s, C.A., K e l l e y, M.C., C a r p e n t e r, L.A., and H o l z-w o r t h, R.H. Evidence for a magnetospheric effect on midlatitude electric fields.–J. Geophys. Res., 83(A9): pp. 4397–4399 (1978).

168. G ö t z–P e t e r B o t h e. The influence of O^+ and NO^+ ions on the structure of the ionospheric F-region.–J. Atmos. Terr. Phys., 36(9): pp. 1537–1545 (1974).

169. G r o v e s, G.V. Seasonal and latitudinal models of atmospheric temperatures, pressure and density, 25 to 110 km. Air Force Cambridge Labs., L.G. Harnison Field, Mass., 1970.

170. H a l l, L.A., H i g g e n s, J.E., C h a g n o n, C.W., and H i n t e r e g g e r, H.E. Solar-cycle variation of extreme ultraviolet radiation.–J. Geophys. Res., 74(16): pp. 4181–4183 (1969).

171. H a l l, L.A., and H i n t e r e g g e r, H.E. Solar radiation in the extreme ultra-violet and its variation with solar rotation.–J. Geophys. Res., 75(34): pp. 6959–6965 (1970).

172. H a n s o n, W.B. Electron temperatures in the upper atmosphere.– Space Res. III, pp. 282–302 (1963).

173. H a n s o n, W.B., and P a t t e r s o n, T.N.L. The maintenance of the night-time F-layer.–Planet. Space Sci., 12(10): pp. 979–997 (1964).

174. H a r p e r, R.M. Nighttime meridional neutral winds near 350 km at low to midlatitudes.–J. Atmos. Terr. Phys., 35(11): pp. 2023–2034 (1973).

175. H a r r i s, I., and P r i s t e r, W. International Reference Atmosphere.– COS-PAR, Amsterdam, 1965.

176. H e d i n, A.E., M a y r, H.G., R e b e r, C.A., S p e n s e r, N.W. and C a r i g n a n, G.R. Empirical model of global thermospheric temperature and composition based on data from the OGO–6 quadrupole mass spectrometer.–J. Geophys. Res., 79(1): pp. 215–225 (1974).

177. H e d i n, A.E. et. al. A global thermospheric model based on mass spectrometer and incoherent scatter data MSIS (2. Composition).–J. Geophys. Res., 82(16): pp. 2148–2156 (1977).

178. H e r o u x, L., and H i n t e r e g g e r, H.E. Aeronomical reference spectrum for solar UV below 2000 A.–J. Geophys. Res., 83(11): pp. 5305–5308 (1979).

179. Hinteregger, H.E. The extreme ultraviolet solar spectrum and its variation during a solar cycle.–Ann. Geophys., 26(2): pp. 547–554 (1970).

180. Hinteregger, H.E. EUV fluxes in the solar spectrum below 2000 A.–J. Atmos. Terr. Phys., 38(8): pp. 791–806 (1976).

181. Hinteregger, H.E. Development of solar cycle 21 observed in EUV spectrum and atmospheric absorption.–J. Geophys. Res., 84(A5): pp. 1933–1938 (1979).

182. Ho, M.C., and Moorcroft, D.R. Hydrogen density and proton flux in the topside ionosphere over Arecibo, Puerto Rico from incoherent scatter observations.–Planet. Space Sci., 19(11): pp. 144–1455 (1971).

183. Ho, M.C., and Moorcroft, D.R. Composition, temperatures and electron loss coefficient of the topside ionosphere over Arecibo.–J. Atmos. Terr. Phys., 39(11/12): pp. 1317–1324 (1977).

184. Hoffman, J.H. Daytime midlatitude ion composition measurements.– J. Geophys. Res., 74(26): pp. 6281–6290 (1969).

185. Hoffman, J.H. Ion mass spectrometer on Explorer XXI satellite.–Proc. IEEE, 57(6): pp. 1063–1067 (1969).

186. Ivanov-Kholodny, G.S. On the interpretation of the seasonal variations of electron density in the F2-region of the ionosphere.–Space Res. IV, pp. 525–533 (1964).

187. Ivanov-Kholodny, G.S. Maintenance of the night ionosphere and corpuscular fluxes in the upper atmosphere.–Space Res. V., N. 1: pp. 19–42 (1965).

188. Jacchia, L.G. Revised static models of the thermosphere and exosphere with empirical temperature profiles.–Spec. Rep. N 332, Smithson. Astrophys. Observ., Cambridge, Mass., 1971.

189. Jacchia, L.G. Variations in thermospheric composition: a model based on mass-spectrometer and satellite-drag data.–Spec. Rep. N. 354, Smithson. Astrophys. Observ., Cambridge, Mass., 1973.

190. Jacchia, L.G. Thermospheric temperature, density and composition: new models.–Spec. Rep. N. 374, Smithson. Astrophjys. Observ., Cambridge, Mass., 1977.

191. Jacchia, L.G., and Slowey, J.W. A study of the variations in the thermosphere related to solar activity.–Space Res. XIII, pp. 343–348 (1973).

192. Jacchia, L.G., Slowey, J.W., and Campbell, I.G. An analysis of the solar-activity effects.–Planet. Space Sci., 21(11): p. 1835 (1973).

193. Jain, A.R., Taylor, G.N, and Williams, P.J.S. Maintenance of the F-region at night: in coherent scatter measurements at a midlatitude station.–J. Atmos. Terr. Phys., 35(10): pp. 1717–1736 (1973).

194. Johnsen, R., and Biondi, M.A. Measurements of the $O^+ + N_2$ and $O^+ + O_2$ reaction rates from 300°K to 2eV.–J. Chem. Phys., 59(7): pp. 3504–3509 (1973).

195. Johnsen, R., Brawn, H.L., and Biondi, M.A. Ion-molecule reactions involving N_2^+, N^+, O_2^+ and O^+ ions from 300°K to 1 eV.–J. Chem. Phys., 52(10): pp. 5080–5084 (1970).

196. Johnson, F.S., and Gottlieb, B. Eddy mixing and circulation at ionospheric levels.–Planet. Space Sci., 18(12): pp. 1707–1718 (1970).

197. Keating, G.M. et. al. The distribution of helium and molecular nitrogen in the lower thermosphere as measured by ESRO 4.–Space Res. XVI, pp. 281–288 (1976).

198. Kelley, M.C., Jörgensen, T.S., and Mikkelsen, J.S. Thermospheric wind measurements in the polar region.–J. Atmos. Terr. Phys., 39(2): pp. 211–219 (1977).

199. Kellogg, W.W. Chemical heating above the polar mesopause in winter.–J. Meteorology, 18(3): pp. 373–381 (1961).

200. King, J.W., and Kohl, H. Upper atmospheric winds and ionospheric drifts caused by neutral air pressure gradients.–Nature, 206(4985): pp. 699–701 (1965).

201. King, J.W. and Smith, P.A. The seasonal anomaly in the behaviour of the $F2$-layer critical frequency.–J. Atmos. Terr. Phys., 30(9): pp. 1707–1713 (1968).

202. Kohl, H., and King, J.W. Atmospheric winds between 100 and 200 km and their effects on the ionosphere.–J. Atmos. Terr. Phys., 29(9): pp. 1045–1062 (1967).

203. Kohl, H., King, J.W., and Eccles, D. Some effects of neutral air winds on the ionospheric F-layer.–J. Atmos. Terr. Phys., 30(10): pp. 1733–1744 (1968).

204. Kotadia, K.M., and Almaula, N.R. Hemispherical $F2$-layer differences and the neutral atmosphere.–J. Atmos. Terr. Phys., 40(5): pp. 623–628 (1978).

205. Labitzke, K. The temperature in the upper stratosphere: differences between hemispheres.–J. Geophys. Res., 79(15): pp. 2171–2175 (1974).

206. Lin, S.L., and Bardsley, J.N. Monte Carlo simulation of ion motion in drift tubes.–J. Chem. Phys., 66(2): pp. 435–445 (1977).

207. Lindinger, W., Fehsenfeld, F.C., Schmeltekopf, A.L., and Ferguson, E.E. Temperature dependence of some ionospheric ion neutral reactions from 300–900°K.–J. Geophys. Res., 79(31): pp. 4753–4756 (1974).

208. Lumb, H.M., and Setty, C.S.G.K. The $F2$-layer seasonal anomaly.–Ann. Geophys., 32(3): pp. 243–256 (1976).

209. Mansurov, S.M., Mansurova, L.G., Mansurov, G.S., Mikhnevich, V.V, and Visotsky, A.M. North–south asymmetry of geomagnetic

and tropospheric events.–J. Atmos. Terr. Phys., 36(11): pp. 1957–1962 (1974).

210. Marubashi, K. Escape of the polar-ionospheric plasma into the magnetic tail.–Rep. Ionos. Space Res. Japan, 24(4): pp. 322–346 (1970).

211. Marubashi, K., and Grebowsky, J.M. A model study of diurnal behavior of the ionosphere and protonosphere coupling.–J. Geophys. Res., 81(10): pp. 1700–1706 (1976).

212. Mayer, H.G., Bauer, P., Brinton, H.C., Brace, L.H., and Potter, W.E. Diurnal and seasonal variations in atomic and molecular oxygen inferred from atmosphere Explorer–C.–Geophys. Res. Lett., 3(2): pp. 77–80 (1976).

213. Mayer, H.G., Harris, I., and Spencer, N.W. Thermosphere "temperatures".–J. Geophys. Res., 79(19): 2921–2924 (1974).

214. Mayr, H.G., and Mahajan, K.K. Seasonal variation in the $F2$ region.–J. Geophys. REs., 76(4): pp. 1017–1027 (1971).

215. McFarland, M., Albritton, D.L., Fehsenfeld, F.C., Ferguson, E.E., and Schmeltekopf, A.L. Flow-drift technique for ion mobility and ion-molecule reaction rate constant measurements. II. Positive ion reactions of N^+, O^+, and N_2^+ with O_2 and O^+ with N_2 from thermal to 2 eV.–J. Chem. Phys. 59(12): pp. 6620–6628 (1973).

216. McNeal, R.J., Whitson, M.E., Jr., and Cook, G.R. Temperature dependence of the quenching of vibrationally excited nitrogen by atomic oxygen.–J. Geophys. Res., 79(10): pp. 1527–1531 (1974).

217. Mentzoni, M.H. and Row, R.V. Rotational excitation and electron relaxation in nitrogen.–Phys. Rev., 130(6): pp. 2312–2316 (1963).

218. Moffett, R.J., Murphy, J.A. Coupling between the F-region and protonosphere. Numerical solution of the time-dependent equations.–Planet. Space Sci., 21(1): pp. 43–52 (1973).

219. Muldrew, D.B. F-layer ionization troughs deduced from Alouette data.–J. Geophys. Res., 70(11): pp. 2635–2650 (1965).

220. Murphy, J.A., Bailey, G.J., and Moffett, R.J. Calculated variations of O^+ and H^+ at midlatitudes–I. Protonospheric replenishment and F-region behaviour at sunspot minimum.–J. Atmos. Terr. Phys., 38(4): pp. 351–364 (1976).

221. Newton, G.P., and Waluer, J.C.G. Electron density decrease in SAR areas resulting from vibrationally excited nitrogen.–J. Geophys. Res., 80(10): pp. 1325–1327 (1975).

222. Newton, G.P., Waluer, J.C.G., Meijer, P.H.E. Vibrationally excited nitrogen in stable auroral red arcs and effect on ionospheric recombination.–J. Geophys. Res., 79(25): pp. 3807–3818 (1974).

223. Nicolet, M., and Swider, W. The ionospheric conditions.–Planet. Space Sci., 11(12): pp. 1459–1484 (1963).

224. Nisbet, J.S. On the construction and use of the Penn. State MK1 ionospheric model.–Ionospheric Res. Lab., The Penn. State Univ., Sci. Rep., N. 355, 98 pp. (1970).

225. Nisbet, J.S. Models of the ionosphere.–Ionosphere Res. Lab., The Penn. State Univ., pp. 245–258 (1974).

226. Norton, R.B., Van Zandt, T.E., and Denison, J.S. A model of the atmosphere and ionosphere in the E and $E1$ regions.– Proceedings of the International Conference in the Ionosphere, The Physical Society, London, p. 26 (1963).

227. Noxon, J.F., and Johanson, A.E. Changes in thermospheric molecular oxygen abundance inferred from twilight 6300A airglow.–Planet. Space Sci., 20(12): pp. 2125–2151 (1972).

228. Offermann, D. Composition variations in the lower thermosphere.– J. Geophys. Res., 79(38): pp. 4281–4293 (1974).

229. Oppenheimer, M., Dalgarno, A., and Trebino, F.P. Daytime chemistry of NO^+ from atmosphere Explorer–C measurements.–J. Geophys. Res., 82(1): pp. 191–194 (1977).

230. Park, C.G. Whistler observations of the interchange of ionization between the ionosphere and the protonosphere.–J. Geophys. Res., 75(22): pp. 4249–4260 (1970).

231. Park, C.G., and Banks, P.M. Influence of thermal plasma flow on the daytime $F2$-layer.–J. Geophys. Res., 80(19): pp. 2819–2823 (1975).

232. Paul, A.K. Temporal and spatial distribution of the spectral components of f_0F2.–J. Atmos. Terr. Phys., 40(2): pp. 135–144 (1978).

233. Piddington, J.H. Ionospheric and magnetospheric anomalies and disturbances.–Planet. Space Sci., 12(6): pp. 553–566 (1964).

234. Prölss, G.W., and von Zahn, V. Esro–r gaz analyzer results. 2. Direct measurements of changes in the neutral composition during an ionospheric storm.–J. Geophys. Res., 79(16): pp. 2535–2539 (1974).

235. Rastogi, R.G. Asymmetry between the $F2$-region of the ionosphere in the northern and southern hemispheres.–J. Geophys. Res., 65(3): pp. 857–858 (1960).

236. Rawer, K., Emmenegger, G., and Schmidtke, G. Statistical analysis of new solar activity measures deduced from satellite EUV–spectra, and classical activity inducers.–XX Plenary meeting of COSPAR, Tel Aviv, Israel, p. 309 (1977).

237. Rawer, K., Ramakrishnan, S., and Bilitza, D. International Ref. Ionosphere, COSPAR, 1975 (1978).

238. Reber, C.A., and Hays, P.B. Thermospheric wind effects on the distribution of helium and argon in the earth's upper atmosphere.–J. Geophys. Res., 78(16): pp. 2977–2991 (1973).

239. Reber, C.A., Hedin, A.E. and Chandra, S. Equatorial phenomena in neutral thermospheric composition.–J. Atmos. Terr. Phys., 35(6): pp. 1223–1228 (1973).

240. Rishbeth, H. Further analogue studies of the ionospheric F-layer.–Proc. Phys. Soc., 81(Pt. 1, N. 519): pp. 65–77 (1963).

241. Rishbeth, H. A time varying model of the ionospheric F2-layer.–J. Atmos. Terr. Phys., 26(6): pp. 657–686 (1964).

242. Rishbeth, H. The effect of winds on the ionospheric F2-peak.–J. Atmos. Terr. Phys., 29(3): pp. 225–238 (1967).

243. Rishbeth, H. The effect of winds on the ionospheric F2-peak.II.–J. Atmos. Terr. Phys., 30(1): pp. 63–71 (1968).

244. Rishbeth, H. Thermospheric winds and the F-region: a review.–J. Atmos. Terr. Phys., 34(1): pp. 1–47 (1972).

245. Rishbeth, H. Drifts and winds in the polar F-region.–J. Atmos. Terr. Phys., 29(1): pp. 111–116 (1977).

246. Rishbeth, H. and Barron, D.W. Equilibrium electron distributions in the ionospheric F2-layer.–J. Atmos. Terr. Phys., 18(2–3): pp. 234–252 (1960).

247. Rishbeth, H., and Kelley, D.M. Maps of the vertical F-layer drifts caused by horizontal winds at mid-latitudes.–J. Atmos. Terr. Phys., 33(4): pp. 539–545 (1971).

248. Rishbeth, H., Moffett, R.J., and Bailey, G.J. Continuity of air motion in the midlatitude thermosphere.–J. Atmos. Terr. Phys., 31(8): pp. 1035–1047 (1969).

249. Rishbeth, H., and Setty, C.S.G.K. The F-layer at sunrise.–J. Atmos. Terr. Phys., 20(4): pp. 263–276 (1961).

250. Roble, R.G. The calculated and observed diurnal variation of the ionosphere over Millstone–Hill on 23–24 March 1970.–Planet. Space Sci., 23(7): pp. 1017–1033 (1975).

251. Roper, P.W., and Baxter, A.J. The effect of auroral input on neutral and ion drifts in the thermosphere.–J. Atmos. Terr. Phys., 40(5): pp. 585–599 (1978).

252. Rothwell, P. Diffusion of ions between F–layers at magnetic conjugate points.–Inter. Conf. on Ionosphere, London, 1962, London, 1963, pp. 217–221.

253. Rüster, R., and King, J.W. Atmospheric composition changes and the F2-layer seasonal anomaly.–J. Atmos. Terr. Phys., 35(7): pp. 1317–1322 (1973).

254. Rüster, R., and King, J.K. Negative ionospheric storms caused by thermospheric winds.–J. Atmos. Terr. Phys., 38(6): pp. 593–598 (1976).

255. Rutherford, J.A., and Vroom, D.A. Effect of metastable $O^+(^2D)$ on

reactions of O^+ with nitrogen molecules.–J. Chem. Phys., 55(12): pp. 5622–5624 (1971).

256. Salah, J.E., Evans, J.V., Alcayde, D., and Bauer, P. Comparison of exospheric temperatures at Millstone–Hill and St. Santin.–Ann. Geophys., 32(1): pp. 257–266 (1976).

257. Salah, J.E., Evans, J.V. and Wand, R.N. Seasonal variations in the thermosphere above Millstone–Hill.–Radio Sci., 9(2): pp. 231–238 (1974).

258. Salah, J.E., and Holt, J.M. Midlatitude thermospheric winds from incoherent scatter radar and theory.–Radio Sci., 9(2): pp. 301–313 (1974).

259. Schmeltekopf, A.L., Ferguson, E.E. and Fehsenfeld, F.C. Afterglow studies of the reactions of He^+, $He(2^3S)$, and O^+ with vibrationally excited N_2.–J. Chem. Phys., 8(7): pp. 2966–2973 (1968).

260. Schmeltekopf, A.L. Ferguson, E.E., Fehsenfeld, F.C. and Cilman, G.I. Reaction of atomic oxygen ions with vibrationally excited nitrogen molecules.–Planet. Space Sci., 15(3): pp. 401–406 (1967).

261. Schmidtke, G. EUV indices for solar-terrestrial relations.– Geophys. Res. Lett., 3(10): pp. 573–576 (1976).

262. Schmidtke, G. Todays knowledge of the solar EUV output and the future needs for more accurate measurements for aeronomy.–Planet. Space Sci., 26(4): pp. 347–353 (1978).

263. Schmidtke, G., Rawer, K., Botzek, H., Norbert, D., and Holzer, K. Solar EUV photon fluxes measured aboard AEROS–A.–J. Geophys. Res., 82(16): pp. 2423–2427 (1977).

264. Schmidtke, G., Rawer, K., Fischer, W., and Rebstock, C. Absolute EUV photon fluxes of aeronomic interest.–Space Res., XV, pp. 345–349 (1975).

265. Schunk, R.W., and Hays, P.B. Theoretical N_2 vibrational distribution in an aurora.–Planet. Space Sci., 21(2): pp. 159–163 (1973).

266. Schunk, R.W., and Walker, J.C.G. Minor ion diffusion in the $F2$-region of the ionosphere.–Planet. Space Sci., 18(9): pp. 1319–1334 (1970).

267. Schunk, R.W., and Walker, J.C.G. Ambipolar diffusion in the $F1$-region of the ionosphere.–Planet. Space Sci., 21(3): pp. 526–528 (1973).

268. Scialom, G. Neutral composition in the lower thermosphere.– Radio Sci., 9(2): pp. 253–262 (1974).

269. Sinha, A.K., and Chandra, S. Seasonal and magnetic storm related changes in the thermosphere induced by eddy mixing.–J. Atmos. Terr. Phys., 36(11): pp. 2055–2066 (1974).

270. Spenner, K. Quiet and disturbed electron temperature and density at different latitudes during daytime.–Space. Res. XV, pp. 363–368 (1975).

271. St.-Maurice, J.P., and Torr, D.G. Non thermal rate coefficients in the ionosphere: the reactions of O^+ with N_2, O_2 and NO.–J. Geophys. Res., 83(3): pp. 969–977 (1978).

272. Sterling, D.L. Influence of electromagnetic drifts and neutral air winds on some features of the F-region.–Radio Sci., 4(11): pp. 1005–1023 (1969).

273. Straus, J.M., Creekmore, S.P., Harris, R.M., and Ching, B.K. Effects of heating at high latitudes on global thermospheric dynamics.–J. Atmos. Terr. Phys., 37(12): pp. 1545–1554 (1975).

274. Stroble, D.F., and McElroy, M.B. The F2-layer at middle latitudes.– Planet. Space Sci., 18(8): pp. 1181–1202 (1970).

275. Stubbe, P. Theory of the nighttime F-layer.–J. Atmos. Terr. Phys., 30(2): pp. 243–263 (1968).

276. Stubbe, P. Simultaneous solution of the time dependent coupled continuity equations heat conduction equations, and equation of motion for a system consisting of a neutral gas, an electron gas, and a four-component ion gas.–J. Atmos. Terr. Phys., 32(5): pp. 865–903 (1970).

277. Stubbe, P. Ionospheric Research, Sci. Rep. 418, Penn. State University, p. 156 (1973).

278. Stubbe, P. The effect of neutral winds on the seasonal F-region variation.–J. Atmos. Terr. Phys., 37(4): pp. 675–680 (1975).

279. Stubbe, P., and Varnum, W.S. Electron energy transfer rates in the ionosphere.–Planet. Space Sci., 20(8): pp. 1121–1126 (1972).

280. Swartz, W.E., and Nisbet, J.S. Revised calculations of F-region ambient electron heating by photoelectrons.–J. Geophys. Res., 77(31): pp. 6259–6261 (1972).

281. Taylor, G.N., and McPherson, P.H. Diurnal and seasonal variations of exospheric heat flux at a midlatitude station.–J. Atmos. Terr. Phys., 36(7): pp. 1135–1146 (1974).

282. Taylor, G.N., Risk, R.J. Empirical relationships between F-region electron density and temperature at Malvern.–J. Atmos. Terr. Phys., 36(8): pp. 1427–1430 (1(1974).

283. Thomas, L. Electron density distribution in the day time F2-layer and their dependence on neutral gas, ion and electron temperatures.–J. Geophys. Res., 71(5): pp. 1357–1366 (1966).

284. Thomas, G.R., and Venables, F.H. The effect of diurnal temperature changes on the F2-layer.–J. Atmos. Terr. Phys., 29(6): pp. 621–640 (1967).

285. Thuillier, G., Falin, J.L., and Barlier, F. Global experimental model of the exospheric temperature using optical and incoherent scatter measurements.–J. Atmos. Terr. Phys., 39(9/10): pp. 1195–1202 (1977).

286. Timothy, A.F., Timothy, J.G. Long-term intensity variations in the solar Helium II Lyman-alpha line.–J. Geophys. Res., 75(34): pp. 6950–6958 (1970).

287. Titheridge, J.E. The stab thickness of midlatitude ionosphere.– Planet. Space Sci., 21(10): pp. 1775–1793 (1973).

288. Torr, M.R., St.–Maurice, J.P. The rate coefficient for the $O^+ + N_2$ reaction in the ionosphere.–J. Geophys. Res., 82(22): pp. 3287–3294 (1977).

289. Torr, M.R., Torr, D.G. The seasonal behavior of the F2-layer of the ionosphere.–J. Atmos. Terr. Phys., 35(12): pp. 2237–2251 (1973).

290. Tyagi, T.R. Electron content and its variation over Lindau.–J. Atmos. Terr. Phys., 36(3): pp. 475–487 (1974).

291. Van Zandt, T.E., and Tinsley, B.A. Theory of F-region recombination processes.–Ann. Geophys., 30(1): pp. 21–33 (1974).

292. Vassuer G. Dynamics of the F-region observed with Thomson scatter. II. Influence of neutral air winds on the ionospheric F-region.–J. Atmos. Terr. Phys., 32(5): pp. 775–787 (1970).

293. Vichland, L.A., and Mason, E.A. Statistical-mechanical theory of gaseous ion-molecule reactions in an electrostatic field.–J. Chem. Phys., 66(2): pp. 422–434 (1978).

294. Vickrey, J., Swartz, W.E., and Farley, D.T. Ion transport in the topside ionosphere.–Inter. Symp. on Radio Waves and the Ionosphere. Helsinki, p. 70 (1978).

295. Von Zahn, U. Neutral air density and composition at 150 kilometers.–J. Geophys. Res., 75(28): pp. 5517–5527 (1970).

296. Von Zahn, U., Frick, K.H., and Trinks, H. ESRO–4 gas analyzer results. First observation of the summer argon bulge.–J. Geophys. Res., 78(31): pp. 1560–1562 (1973).

297. Von Zahn, U., Köhnlein, W., Frick, K.H., Laux, U., Trinks, H., and Volland, H. ESRO–4 model of global thermospheric composition and temperatures during times of low solar activity.– Geophys. Res. Letters., 4(1): pp. 33–36 (1977).

298. Waldman, H. The specification of distributed boundary conditions in numerical simulation of the ionosphere.–J. Atmos. Terr. Phys., 35(12): pp. 2205–2215 (1973).

299. Walker, J.C.G., Stolarski, R.S., and Nagy, A.F. The vibrational temperature of molecular nitrogen in the thermosphere.– Ann. Geophys., 25(4): pp. 831–839 (1969).

300. Woodgate, B.E., Knight, D.E., Uribe, R., Sheather, P., Bowles, J., and Nettleship, R. Extreme ultraviolet line intensities from the Sun.–Proc. Roy. Soc. Cont., A332(1590): pp. 291–309 (1973).

301. Woodman, R.F. Vertical drift velocities and east–west electric fields as the magnetic equator.–J. Geophys. Res., 75(31): pp. 6249–6259 (1970).

302. Yonezawa, T., and Arima, T. On the seasonal and non-seasonal annual variations and the semiannual variation in the noon and midnight electron densities of the $F2$-layer in middle latitudes.–J. Radio Res. Labor., 6(25): pp. 293–310 (1959).

303. Yonezawa, T. Maintenance of ionization in the nighttime $F2$-region.–Space. Res., V, pp. 49–60 (1965).

304. Yonezawa, T. Theory of the formation of the ionosphere.–Space Sci. Rev., 5(1): pp. 3–56 (1966).

305. Yonezawa, T. The vertical distribution of ionization at the $F2$-peak related to ionic production and transport processes.– Ann. Geophys., 26(2): pp. 581–588 (1970).

306. Yonezawa, T. The solar-activity and latitudinal characteristics of the seasonal, non-seasonal and semi-annual variations in the peak electron densities of the $F2$-layer at noon and at midnight in middle and low altitudes.–J. Atmos. Terr. Phys., 33(6): pp. 889–907 (1971).

Subject Index